Current Topics in Microbiology and Immunology

160

Editors

Retrovirus Infections of the Nervous System

Current and Future Perspectives

Edited by
M. B. A. Oldstone and H. Koprowski

With 16 Figures

Springer-Verlag
Berlin Heidelberg New York
London Paris Tokyo Hong Kong

Michael B. A. Oldstone, M.D.
Dept. of Neuropharmacology, Scripps Clinic and Research Foundation, 10666 N. Torrey Pines Road, La Jolla, CA 92037, USA

Hilary Koprowski, M.D., Direktor
The Wistar Institute, 36th Street at Spruce, Philadelphia, PA 19104, USA

ISBN-13:978-3-642-75269-8 e-ISBN-13:978-3-642-75267-4
DOI: 10.1007/978-3-642-75267-4

Softcover reprint of the hardcover 1st edition 1990

Library of Congress Catalog Card Number 15-12910

Phototypesetting by Thomson Press (India) Ltd, New Delhi

2123/3020-543210 – Printed on acid-free paper

Preface

Although retroviruses have long been associated with a variety of animal diseases, active research in the field of human retroviruses dates from the discovery of human immunodeficiency virus (HIV) in association with acquired immunodeficiency syndrome (AIDS). The enormous research efforts in this field have been directed toward understanding the nature of the virus and toward its elimination through preventive vaccination and the cure of the disease.

Human T-cell leukemia virus (HTLV-1) was the first member of the human retrovirus family to be discovered. It was implicated as the cause of adult T-cell leukemia (ATL) even before the association of HIV and AIDS was established. Research on HTLV-1 has, however, been lagging behind that of HIV because of the importance of AIDS. Today HTLV-1 and possibly closely related HTLV-2 are associated with a variety of human neurologic diseases, and research activities in this field may show that human retroviruses can cause a variety of human diseases in addition to those affecting the nervous system.

Papers in this volume attempt to acquaint the reader with the present state of research into retrovirus infection and related diseases of the nervous system.

Prominent among diseases of experimental and domestic animals caused by retroviruses are demyelinative disorders, arthritic and rheumatic illnesses, and progressive motor neuron disorders. The wild mouse ecotropic virus causes a disease that is discussed in the chapters by GARDNER and PORTIS. GARDNER first described the disease in wild ("outbred") mouse populations and he has contributed to defining the susceptibility/resistance genes that control this infection. PORTIS has been studying the disorder in inbred populations and mapping the viral gene(s) involved. The study of wild mouse ecotropic virus demonstrates how a retrovirus functions in a genetically defined and manipulable experimental animal to cause slow, progressive disease and points up the similarities between this disorder and the human disease, amyotropic lateral sclcrosis.

To define viral genes and products in murine retrovirus-neurologic disease, WONG has developed temperature-sensitive mutants, variants, and revertants of Moloney murine leukemia virus and studied their ability to cause neurologic disease. Although MCGUIRE deals not with retroviruses but with lentiviruses infection, the caprine disease of the CNS provides a good model for the study of human retrovirus infection. The chapter by LACKNER et al. reports the study of retrovirus infection in nonhuman primates and describes the pathogenesis of the disease in the species nearest to the human host.

In the section on human retrovirus infections, MCFARLIN and KOPROWSKI deal with the presence of HTLV-1 or HTLV-1-related virus in tissues of patients with chronic progressive myelopathies such as tropical spastic paraparesis/HTLV-1-associated myelopathy syndrome and multiple sclerosis. Attempts to produce an HTLV-1 transgenic mouse model are described by NERENBERG. The three other chapters in this section deal with infections of the nervous system by HIV. GNANN and OLDSTONE describe the precise diagnostic tools necessary to distinguish among infections caused by the variety of HIV strains. Attempts to identify a "neurotropic" strain of HIV are described by CHENG-MEYER and LEVY. Finally, the pathogenesis of nervous system infection by HIV is dealt with by WILEY and NELSON.

The editors realize that this volume covers only a fraction of research activity in the field of human retroviruses and neurologic diseases. They hope, however, that this publication will promote better understanding of the role of these viruses in a variety of human and animal disease and that this will increase the interest in research in this important area.

Michael B. A. Oldstone
La Jolla, CA

Hilary Koprowski
Philadelphia, PA

List of Contents

List of Contributors

(Their addresses can be found at the beginning of their respective chapters)

CHEEVERS, W. P.
CHENG-MAYER, C.
GARDNER, M. P.
GNANN, J. W.
KNOWLES, D. P.
KOPROWSKI, H.
LACKNER, A. A.
LEVY, J. A.
LOWENSTINE, L. G.
MARX, P. A.
MCFARLIN, D. E.
MCGUIRE, T. C.
NELSON, J. A.
NERENBERG, M. J.
OLDSTONE, M. B. A.
O'ROURKE, K. I.
PORTIS, J. L.
WILEY, C. A.

I. Animal Retrovirus Infections of the Nervous System

Genetic Resistance to a Retroviral Neurologic Disease in Wild Mice

M. B. GARDNER

1 Introduction

The only known example of a naturally occurring oncoviral neurologic disease in animals is the fatal hind leg paralysis described in an aging population of wild mice (*Mus musculus*) from an isolated squab farm near Lake Casitas (LC) in southern California (OFFICER et al. 1973; GARDNER et al. 1973). The disease, called spongiform polioencephalomyelopathy, is caused by an infectious (ecotropic) murine leukemia virus (MuLV) and is characterized by a noninflammatory, nonimmunogenic, direct retrovirus injury to anterior horn neurons in the lumbosacral spinal cord. The observation that this virus and its associated paralysis are detected in only a small minority (10%) of the LC mice is explained by the segregation in this outbred population of a dominant ecotropic-MuLV restriction gene, initially called *Akrv-1* and now called *FV-4*. In this chapter I briefly summarize the natural history of this retroviral neurologic disease, for which several recent reviews are available (GARDNER 1978; GARDNER and RASHEED 1982; GARDNER 1985), and cover more recent information on the discovery, structure, and function of this MuLV restriction gene.

2 Natural History

The major natural history features of the MuLV neurologic disease in LC wild mice are listed in Table 1. There are two classes of infectious MuLV present in LC mice; one class is called amphotropic because of a broad *in vitro* host range for

Department of Medical Pathology, University of California, Davis, California 95616, USA

Table 1. Natural history features of MuLV in LC wild mice

1.	Two classes of infectious MuLV are present, amphotropic virus in normal, lymphomatous, and paralyzed LC mice, and ecotropic virus mainly in paralyzed LC mice.
2.	Both classes of MuLV are exogenous and transmitted by maternal congenital infection leading to lifelong viremia and specific viral immunologic tolerance. General immunity is intact.
3.	Experimental transmission of the LC MuLVs, including molecularly cloned virus, shows that both classes of MuLV induce B lymphomas whereas only the ecotropic MuLV induces paralysis.
4.	Ecotropic MuLV may arise by recombination of amphotropic MuLV with endogenous MuLV-related sequences in feral and laboratory mice. Recombination with endogenous MuLV virogenes may occur in the spleen during lymphomogenesis but is not involved with paralytogenesis in the CNS.
5.	About 15% of LC mice in their natural habitat escape congenital infection because their mothers are nonviremic; they remain persistently nonviremic, free of paralysis and lymphoma, and show no detectable autogenous immunity to MuLV.

murine and nonmurine cells, and the other class is called ecotropic because its host range is restricted to murine cells. Both classes of MuLV are related to but distinct from the infectious MuLV of laboratory mouse strains. Amphotropic MuLV has been found only in wild mice and may give rise to ecotropic MuLV by recombination with endogenous MuLV virogenes (RASHEED et al. 1976, 1983). By far the most prevalent MuLV is the amphotropic class found in about 85% of LC mice from shortly after birth and throughout life (GARDNER et al. 1976a, b, 1978). Amphotropic virus is only weakly lymphomogenic, inducing ≤ 20% lymphomas after about 12 months in susceptible laboratory mice. It is not paralytogenic. By contrast, the ecotropic virus is isolated together with amphotropic virus mainly from paralyzed LC mice; the ecotropic virus alone, including molecular cloned virus (JOLICOEUR et al. 1983), induces 70%–100% paralysis in 1–8 months in newborn susceptible lab mice and is also lymphomogenic if the animals live long enough (≥ 8 months). The lymphomas, naturally occurring or experimentally induced by either class of LC MuLV, are B-cell origin and arise in the spleen (BRYANT et al. 1981). Recombination of wild mouse MuLV with endogenous MuLV virogenes may occur in the spleen during leukemogenesis in laboratory mice (HOFFMAN et al. 1981; LANGDON et al. 1983; RASSART et al. 1986), but recombination is not involved during paralytogenesis (RASHEED et al. 1983; OLDSTONE et al. 1983). The ecotropic viral determinants of paralysis lie within several domains in the envelope gene (RASSART et al. 1986; DESGROSEILLERS and JOLICOEUR 1984), whereas the determinants of leukemogenicity in both amphotropic and ecotropic virus map to multiple regions of the genome (DESGROSEILLERS et al. 1984; JOLICOEUR and DESGROSEILLERS 1985).

The LC MuLVs are solely exogenous (BARBACID et al. 1979; O'NEILL et al. 1987) and transmitted primarily by maternal congenital infection, mainly via milk (GARDNER et al. 1979). Horizontal or casual transmission does not occur. Venereal transmission may occur under experimental circumstances (GARDNER et al. 1979; PORTIS et al. 1987). Infection with MuLV in the newborn period is accompanied by specific immunologic tolerance (KLEMENT et al. 1976), systemic spread, and lifelong viremia with high levels of virus (GARDNER 1976). General immunity, however, is intact. Measures to reduce the level of MuLV early in life, such as foster nursing on nonviremic mothers, splenectomy, or passive immunization, reduce or eliminate the later development of paralysis and lymphoma (HENDERSON et al. 1974). Neither disease develops in the 15% of nonviremic LC mice that escape congenital MuLV infection because their mothers are nonviremic (GARDNER et al. 1980a).

The major pathologic features of the naturally occurring and experimentally induced neurologic disease are a noninflammatory spongiform change accompanied by loss of anterior horn neurons and reactive gliosis (ANDREWS and GARDNER 1974; OLDSTONE et al. 1980; BROOKS et al. 1980). These changes are most severe in the lower spinal cord, although in experimentally infected newborn laboratory mice the changes can occur throughout the spinal cord and into the brain stem and cerebellum. However, the cerebral cortices are never involved. This lower motor neuron disease is accompanied by fasciculations and flaccid paralysis of the lower limbs and motor unit atrophy of skeletal muscle fibers. Numerous type C virus particles are seen in the capillary endothelia and extracellular space of the affected spinal cords. Budding type C particles are also seen in capillary endothelia and, occasionally, in anterior horn neurons. A few anterior horn neurons contain aberrant, abortively replicating type C particles in vacuoles or distended rough endoplasmic reticulum. Virus antigen is detected in endothelia, anterior horn neurons, and, possibly, some glial cells (OLDSTONE et al. 1980).

The pathogenesis of this degenerative "slow virus" neurologic disease thus appears to be primarily one of direct viral injury to neurons without an associated immunologic or inflammatory host response. Ecotropic MuLV, acquired at birth, or by recombination with amphotropic MuLV after birth, replicates mainly in the spleen and spreads early in life to the CNS as cell-free virus in the bloodstream. Further virus replication then takes place in the CNS, primarily in endothelial cells and, after reaching a critical threshold, binding to and entry into anterior horn neurons occurs. Abortive replication of virus in these nondividing neuronal cells leads to an accumulation of viral products in the cytoplasm which damages cell membranes, causes loss of the normal selective permeability, and leads to cell swelling and death. Damage to neuronal, glial, and endothelial cell membranes from extracellular viral products, e.g., envelope glycoprotein, may also contribute to the spongiform degeneration.

In summary, the natural history features of MuLV in lymphoma and paralysis-prone LC wild mice are quite different from those in lymphoma-prone AKR inbred laboratory mice. The wild mouse MuLVs most closely resemble the exogenous Friend, Moloney, and Rauscher MuLV strains of laboratory mice (BARBACID et al. 1979). Perhaps this explains why a similar neurologic disease has been described in mice and rats following experimental inoculation of mutant strains of Moloney and Friend MuLV, respectively (WONG et al. 1983; KAI and FURUTA 1984). Similarly, variants of amphotropic virus have been isolated from cultures of Moloney and Rauscher MuLV-induced lymphomas in laboratory mice (CHANG et al. 1988). In many respects, especially the exogenous route of virus transmission by maternal congenital infection, the wild mouse MuLV model has proved much more accurate than the laboratory mouse model in predicting the natural history features of the first human leukemia virus (HTLV-1) discovered almost a decade later. The value of this wild mouse model for human leukemia has recently been discussed (GARDNER 1987).

3 Discovery and Characterization of Akvr-1/FV-4

The discovery of the *Akvr-1*^{R} gene occurred serendipitously while crossbreeding LC wild mice with AKR inbred mice to see what fate would befall the progeny of such MuLV-laden lymphoma-prone parental mouse strains (GARDNER et al. 1980b). Three patterns emerged in analyzing the F_1 progeny for MuLV viremia at 2 months of age: all of the F_1 progeny from individual matings were viremic, all were nonviremic, or about 50% were viremic. The viremia was always due to AKR ecotropic virus because only LC males or nonviremic LC females were bred to the AKR mice and we had already determined that the LC MuLV is transmitted in nature only by viremic LC females (GARDNER et al. 1979). This pattern of MuLV viremia in the F_1 offspring strongly suggested the segregation, in LC mice, of a dominant gene capable of strongly blocking the expression of the ecotropic MuLV inherited from the AKR parent. Confirmation was obtained by measuring viremia in F_1 backcrosses to AKR and in F_2 mice. As expected from the effect of a single dominant gene, approximately 50% and 25% viremia, respectively, was observed in these crosses. This gene was called *Akvr-1*^{R} because of its restrictive or blocking effect on expression or replication of the AKR ecotropic MuLV (*Akv*). Properties of this MuLV restriction gene are summarized in Table 2. The dominant *Akv* restricion allele was called *R* and the recessive *Akv* susceptibility allele was first called *r* then *S*. LC mice whose AKR F_1 progeny were 100% nonviremic are homozygous for this resistance allele (*Akvr-1*^{RR}). LC mice whose AKR F_1 progeny were 50% viremic are heterozygous (*Akvr-1*^{RS}) and LC mice whose AKR F_1 progeny were 100% viremic are homozygous for the

Table 2. Properties of *Akvr-1*^{R}/*FV-4*^{R}

1. Dominant gene, segregating in LC wild mice, and present in about 72% of this population.
2. Strongly blocks replication of all ecotropic MuLVs at the cellular level, in vivo and in vitro.
3. Allelic with and sequence identical to *FV-4*^{R} on chromosome 12 in Japanese wild mice and the derived FRG inbred mouse strain.
4. Represents a defective endogenous, ecotropic MuLV-related provirus encoding an envelope glycoprotein (gp70) that interferes at the cell surface with entry of ecotropic viruses.
5. Segregation of this gene in LC wild mice controls susceptibility or resistance to the ecotropic MuLV and associated paralysis and lymphoma indigenous in this outbred population.

recessive allele (*Akvr-1*^{SS}). Concordant resistance to experimental infection with Friend and Moloney ecotropic MuLV was also observed. However, LC males, homozygous or heterozygous for *Akvr-1*^{R}, were still viremic with amphotropic MuLV, indicating that this restriction allele does not block replication of amphotropic virus. Crosses of LC^{RR} mice with NIH Swiss laboratory mice experimentally infected with LC ecotropic virus or with strains of inbred mice bearing endogenous AKR-related MuLV (e.g., PL/J, C_3H/Fg, C58) confirmed the dominant restriction effect of the *Akvr-1*^{R} allele on all of these ecotropic viruses (GARDNER, unpublished data). F_1 progeny of matings between AKR female and LC males homozygous for the resistance allele (*Akvr-1*^{RR}) show that the MuLV restriction effect was long lasting ($\geqslant$18 months) and associated with prevention of lymphoma.

The same restriction effect was observed in vitro (RASHEED and GARDNER 1983). Cultured LC^{RR} mouse fibroblasts and hematopoietic cells were resistant to infection with ecotropic MuLVs including those of LC mouse origin whereas these cells were fully permissive to infection with amphotropic virus. However, the block to infection in vitro, especially in cells of heterozygous (*Akvr-1*^{RS}) mice, was not as strong as that observed in vivo. It is clear that *Akvr-1*^{R} is distinct from other MuLV restriction genes such as FV-1, FV-2, and FV-3 which have been described in laboratory mice. None of these other genes acts as strongly as *Akrv-1*^{R}. Furthermore all LC mice are homozygous for *FV-1*^{N} and therefore permissive at this locus for replication of their ecotropic and amphotropic MuLVs which are all N-tropic for mouse cells.

The observed frequency of the *Akvr-1*^{R} allele (47%) and the *Akvr-1*genotype frequencies observed in LC mice did not vary significantly from expectations of the Hardy-Weinberg equilibrium, indicating that segregation of this allele follows the rules of Mendelian genetics. Thus, the probable frequency of LC mice that contain at least one *Akvr-1*^{R} allele is 72%. All these mice are, therefore, resistant to ecotropic MuLV and this resistant to the eventual development of the associated paralysis or lymphoma. They are not, however, totally resistant to lymphoma, because they may still be infected with amphotropic MuLV.

Prior to the discovery of *Akvr-1*^{R} in LC wild mice, a similar gene, called *FV-4*^{R}, was described in Japanese wild mice (*Mus molossinus*); an inbred strain (FRG strain) homozygous for *FV-4*^{R} was derived from these wild mice (ODAKA et al. 1981). *FV-4*^{R} shows the same restriction effect, in vivo and in vitro, on ecotropic MuLV infection as does *Akvr-1*^{R}. *FV-4*^{R} maps to mouse chromosome 12. Crosses between (LC^{RR} × FRG) hybrids and AKR mice failed to produce any viremic offspring, suggesting that these genes are alleles of a single locus or are closely linked genes (O'BRIEN et al. 1983).

A viral interference model was proposed as the mechanism of *FV-4* restriction based on the association of this gene, in uninfectd cells, with a unique cell surface antigen related to ecotropic MuLV envelope glycoprotein (gp70) (IKEDA and ODAKA 1984). The DNA encoding this gp70 was molecularly cloned from *FV-4*^{R} congenic BALB/c mice; it represents a defective MuLV proviral locus containing an ecotropic MuLV *env*-related sequence (KOZAK et al. 1984). Presumably, ecotropic viral interference results from filling of the cell surface receptors with this endogenous gp70. All ecotropic MuLVs use the same cell receptor and thus would be expected to be blocked by this mechanism (SARMA et al. 1967). Precedence for retroviral interference at the cell surface level from receptor blockage with endogenous defective proviral gp70 was first described in the avain leukosis virus model system (PAYNE et al. 1971; ROBINSON et al. 1981).

We made similar observations with the *Akvr*-1[R] gene in LC wild mice (DANDEKAR et al. 1987). As with *FV*-4[R], the *Akvr*-1[R] gene is linked to an ecotropic MuLV-related gp70 which blocks entry of ecotropic MuLV into uninfected target cells. A unique defective endogenous provirus, segregating with *Akvr*-1 resistance among LC feral mice, was molecularly cloned from LC mouse genomic DNA and it shows complete nucleotide homology with the *FV*-4[R]-asociated provirus. This homology also extends to the 3′ host flanking chromosomal DNA. These findings conclusively establish that *Akvr*-1[R] and *FV*-4[R] are the same gene and suggest that a piece of chromosome 12 was probably exchanged in recent times by interbreeding of these distinct feral mouse species. Henceforth, these genes should be considered as members of an allelic series of a single locus, *FV*-4. *Fv*-4[R] has not been found in North American laboratory mice, presumably because of the narrow genetic base from which they were derived (MORSE 1978). However, a similar restriction gene for MCF-recombinant MuLV has been described in the DBA inbred mouse strain (BASSIN et al. 1982). The MCF recombinant MuLV uses a different receptor on mouse cells than the ecotropic virus (REIN 1982). Evolutionary conservation of this locus is probably not favored in nature because the associated diseases in LC mice occur well after breeding age. However, susceptibility or resistance to the natural occurrence of spongiform polioencephalomyelopathy in aging LC wild mice must depend primarily on inheritance of this MuLV-resistance gene segregating in this outbred population. Whether or not a similar genetic resistance applies to other retroviral diseases of animals or humans remains unknown.

References

Andrews JM, Gardner MB (1974) Lower motor neuron degeneration associated with type C RNA virus infection in mice: neuropathological features. J Neuropathol Exp Neurol 33: 285–307

Barbacid M, Robbins KC, Aaronson SA (1979) Wild mouse RNA tumor viruses: a nongenetically transmitted virus group closely related to exogenous leukemia viruses of laboratory mouse strains. J Exp Med 149: 254–266

Bassin RH, Ruscetti S, Iqbal A, Haapala DK, Rein A (1982) Normal DBA mouse cells synthesize a glycoprotein which interferes with MCF virus infection. Virology 123: 139–151

Brooks BR, Swarz Jr, Johnson RT (1980) Spongiform polioencephalomyelopathy caused by a murine retrovirus. I. Pathogenesis of infection in newborn mice. Lab Invest 43: 480–486

Bryant ML, Scott JL, Pak BK, Estes JD, Gardner MB (1981) Immunopathology of natural and experimental lymphomas induced by wild mouse leukemia virus. Am J Pathol 104: 272–282

Chang KSS, Wang L, Gao C (1988) Variants of amphotropic type C retrovirus isolated from cultures of Moloney and Rasucher-MuLV-induced tumors. Int J Cancer 41: 756–761

Dandekar S, Rossitto P, Pickett S, Mockli L, Bradshaw H, Cardiff R, Gardner M (1987) Molecular characterization of the *Akvr-1* restriction gene: a defective endogenous retrovirus-borne gene identical to *Fv-4*^{r}. J Virol 61: 308–314

DesGroseillers L, Jolicoeur P (1984) Mapping the viral sequences conferring leukemogenicity and disease specificity in Moloney and amphotropic murine leukemia viruses. J Virol 52: 448–456

DesGroseillers L, Barrett M, Jolicoeur P (1984) Physical mapping of paralysis-inducing determinant of a wild mouse ecotropic neurotropic retrovirus. J Virol 52: 356–363

Gardner MB (1978) Type C viruses of wild mice: characterization and natural history of amphotropic, ecotropic, and xenotropic MulV. In: Current topics in microbiology and immunology, vol 79, Springer, Bertin Heidelberg New York, pp 215–259

Gardner MB (1985) Retroviral spongiform polioencephalomyelopathy. Rev Infect Dis 7: 99–110

Gardner MB (1987) Naturally occurring leukaemia viruses in wild mice: how good a model for humans? Cancer Surv 6: 55–71

Gardner MB, Rasheed S (1982) Retroviruses in feral mice. Int Rev Exp Pathol 23: 209–267

Gardner MB, Henderson BE, Officer JE, Rongey RW, Parker JC, Oliver C, Estes JD, Huebner RJ (1973) A spontaneous lower motor neuron disease apparently caused by indigenous type-C RNA virus in wild mice. J Natl Cancer Inst 51: 1243–1254

Gardner MB, Henderson BE, Estes JD, Rongey RW, Casagrande J, Pike M, Huebner RJ (1976a) The epidemiology and virology of C-type virus-associated hematological cancers and related diseases in wild mice. Cancer Res 36: 574–581

Gardner MB, Klement V, Rongey RR, McConahey P, Este JD, Huebner RJ (1976b) Type C virus expression in lymphoma-paralysis-prone wild mice. J Natl Cancer Inst 57: 585–590

Gardner MB, Klement V, Rasheed S, Estes JD, Rongey RW, Dougherty MF, Bryant ML (1978) In: Steves JG, Todaro GJ, Fox CF (eds) Persistent MuLV infection in wild mice. Academic, New York, pp 115–131

Gardner MB, Chiri A, Dougherty MF, Casagrande J, Estes JD (1979) Congenital transmission of murine leukemia virus from wild mice prone to the development of lymphoma and paralysis. J Natl Cancer Inst 62: 63–70

Gardner MB, Rasheed S, Estes JD, Casagrande J (1980a) The history of viruses and cancer in wild mice. Cold Spring Harbor Laboratory, New York, pp 971–987

Gardner MB, Rasheed S, Pal BK, Estes JD, O'Brien SJ (1980b) *Akvr*-1, a dominant murine leukemia virus restriction gene, is polymorphic in leukemia-prone wild mice. Proc Natl Acad Sci USA 77: 531–535

Henderson BE, Gardner MB, Gilden RV, Estes JD, Huebner RJ (1974) Prevention of lower limb paralysis by neutralization of type-C RNA virus in wild mice. J Natl Cancer Inst 53: 1091–1092

Hoffman PM, Davidson WF, Ruscetti SK, Chused TM, Morse HC III (1981) Wild mouse ecotropic murine leukemia virus infection of inbred mice: dual-tropic virus expression precedes the onset of paralysis and lymphoma. J Virol 39: 597–602

Ikeda H, Odaka T (1984) A cell membrane "gp70" associated with *Fv-4* gene: immunological characterization and tissue and strain distribution. Virology 133: 65–76

Jolicoeur P, DesGroseillers L (1985) Neurotropic cas-BR-E murine leukemia virus harbors several determinants of leukemogenicity mapping in different regions of the genome. J Virol 56: 639–643

Jolicoeur P, Nicolaiew N, DesGroseillers L, Rassart E (1983) Molecular cloning of infectious viral DNA from ecotropic neurotropic wild mouse retrovirus. J Virol 56: 639–643

Kai K, Furuta T (1984) Isolation of paralysis-inducing murine leukemia viruses from Friend virus passaged in rats. J Virol 50: 970–973

Klement V, Gardner MB, Henderson BE, Ihle JN, Estes JD, Stanley AG, Gilden RV (1976) Inefficient humoral immune response of lymphoma-prone wild mice to persistent leukemia virus infection. J Natl Cancer Inst 57: 1169–1173

Kozak CA, Gromet NJ, Ikeda H, Buckler CE (1984) A unique sequence related to the ecotropic murine leukemia virus is associated with the *Fv-4* resistance gene. Proc Natl Acad Sci USA 81: 834–837

Langdon WY, Hoffman PM, Silver JE, Buckler CE, Hartley JW, Ruscetti SK, Morse HC III (1983) Identification of a spleen focus-forming virus in erythroleukemic mice infected with a wild-mouse ecotropic murine leukemia virus. J Virol 46: 230–238

Morse HC III (1978) Origins of inbred mice. Academic, New York

O'Brien SJ, Berman EJ, Estes JD, Gardner MB (1983) Murine retroviral restriction genes *Fv-4* and *Akvr-1* are alleles of a single locus. J Virol 47: 649–651

Odaka T, Ikeda H, Yoshikura H, Moriwaki K, Suzuki S (1981) *Fv-4*: gene controlling resistance to

NB-tropic Friend murine leukemia virus. Distribution in wild mice, introduction into genetic background of BALB/c mice, and mapping of chromosomes. J Natl Cancer Inst 67: 1123–1127
Officer JE, Tecson N, Estes JD, Fontanilla E, Rongey RW, Gardner MB (1973) Isolation of a neurotropic type C virus. Science 181: 945–947
Oldstone MBA, Jensen F, Dixon FJ, Lampert PW (1980) Pathogenesis of the slow disease of the central nervous system associated with wild mouse virus. II. Role of virus and host gene products. Virology 107: 180–193
Oldstone MBA, Jensen F, Elder J, Dixon FJ, Lampert PW (1983) Pathogenesis of the slow disease of the central nervous system associated with wild mouse virus. III. Role of input virus and MCF recombinants in disease. Virology 128: 154–165
O'Neill RR, Hartley JW, Repaske R, Kozak CA (1987) Amphotropic proviral envelope sequences are absent from the *Mus* germline. J Virol 61: 2225–2231
Payne LN, Poni PK, Weiss RA (1971) A dominant epistatic gene which inhibits cellular susceptibility to RSV (RAV-O). J Gen Virol 13: 455–462
Portis JL, McAtee FJ, Hayes SF (1987) Horizontal transmission of murine retroviruses. J Virol 61: 1037–1044
Rasheed S, Gardner MB (1983) Resistance of fibroblasts and hematopoietic cells to ecotropic murine leukemia virus infection; an *Akvr-1*^{R} gene effect. Int J Cancer 31: 491–496
Rasheed S, Gardner MB, Chan E (1976) Amphotropic host range of naturally occurring wild mouse leukemia viruses. J Virol 19: 13–18
Rasheed S, Gardner MB, Lai MMC (1983) Isolation and characterization of new ecotropic murine leukemia viruses after passage of an amphotropic virus in NIH Swiss mice. Virology 130: 439–451
Rassart E, Nelbach L, Jolicoeur P (1986) The *Cas-Br-E* MuLV: sequencing of the paralytogenic region of its genome and derivation of specific probes to study its origin and the structure of its recombinant genomes in leukemic tissues. J Virol 60: 910–919
Rein A (1982) Interference grouping of murine leukemia viruses: a distinct receptor for the MCF-recombinant viruses in mouse cells. Virology 120: 251–257
Robinson HL, Astrin SM, Senior AM, Salazar FH (1981) Host susceptibility to endogenous viruses: defective glycoprotein-expressing proviruses interfere with infections. J Virol 40: 745–751
Sarma PS, Cheong MP, Hartley JW, Huebner RJ (1967) A viral interference test for mouse leukemia viruses. Virology 33: 180–184
Wong PKY, Soong MM, MacLeod R, Gallick GE, Yuen PH (1983) A group of temperature-sensitive mutants of Moloney leukemia virus which is defective in cleavage of *env* precursor polypeptide in infected cells also induces hind-limb paralysis in newborn CFW/D mice. Virology 125: 513–518

Wild Mouse Retrovirus: Pathogenesis

J. L. PORTIS

1 Introduction

GARDNER and coworkers (GARDNER et al. 1973) discovered in a population of wild mice a neurologic disease manifested primarily by hind limb paralysis. Pathologic studies indicated that the motor neurons as well as glial elements in the spinal cord and hindbrain underwent progressive vacuolar degeneration associated with gliosis which was unaccompanied by evidence of inflammation (GARDNER et al. 1973; ANDREWS and GARDNER 1974; SWARZ et al. 1981). A retrovirus isolated from the CNS of these mice was found to reproduce the disease in laboratory mice (ANDREWS and GARDNER 1974), making this the first example of a purely degenerative disease of the CNS known to be caused by a retrovirus.

The pathogenesis of the disease was initially approached using electron microscopy and classical virologic techniques to identify the sites of virus replication (GARDNER et al. 1973; ANDREWS and GARDNER 1974; BROOKS et al. 1980). Virus was seen both in the CNS (primarily in the hindbrain and spinal cord) and in skeletal muscle. However, the paralysis was associated with neurogenic atrophy of the skeletal muscles, indicating that this was not a primary

Laboratory of Persistent Viral Diseases, National Institute of Allergy and Infectious Diseases, Rocky Mountain Laboratories, Hamilton, Montana 59840, USA

Current Topics in Microbiology and Immunology, Vol. 160

muscle disorder but a disease of the CNS. In the CNS, virus was associated with vascular endothelial and perithelial cells where particles were seen budding from the plasma membrane and accumulating extracellularly (ANDREWS and GARDNER 1974; SWARZ et al. 1981; PITTS et al. 1987). In addition, virus was seen within neurons (GARDNER et al. 1973; ANDREWS and GARDNER 1974; OLDSTONE et al. 1977) and oligodendroglia (OLDSTONE et al. 1977) budding into intracytoplasmic vesicles probably representing endoplasmic reticulum (Fig. 1). Particles in these locations exhibited bizarre shapes, often forming extended tubular structures. Despite this striking observation, however, many of the vacuolar degenerative changes seen in this disease were unassociated with recognizable virus particles (ANDREWS and GARDNER 1974; SWARZ et al. 1981). In this respect, the wild mouse disease resembles the neurologic disease caused by the neurovirulent *ts*1 mutant of Mo-MuLV in which CNS degeneration is unaccompanied by ultrastructural evidence of virus assembly (ZACHARY et al. 1986).

The paralytic disease induced by WM-E does not appear to involve immunopathologic mechanisms for two reasons: (a) there is little evidence until

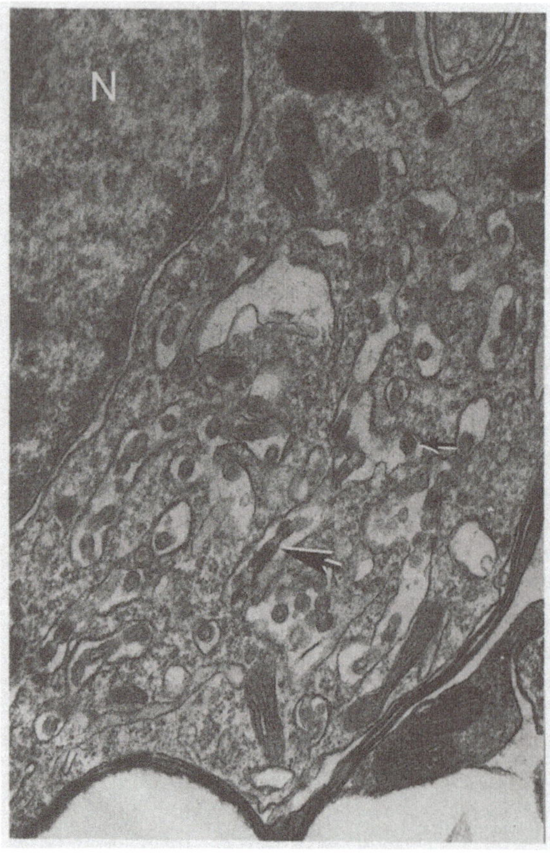

Fig. 1. Electron micrograph of a neuron in the ventral horn of the lumbar spinal cord of a paralyzed 6-month-old IRW mouse neonatally inoculated intraperitoneally with WM-E. This cell is located at the junction of the gray and white matter and illustrates the characteristic intracytoplasmic budding of virus (*arrows*). Virus particles exhibit aberrant forms including tubular structures with varicosities (*large arrow*). The separation of cytoplasm from nuclear membrane is a fixation artifact. *N*, nucleus. Final magnification, ×33000

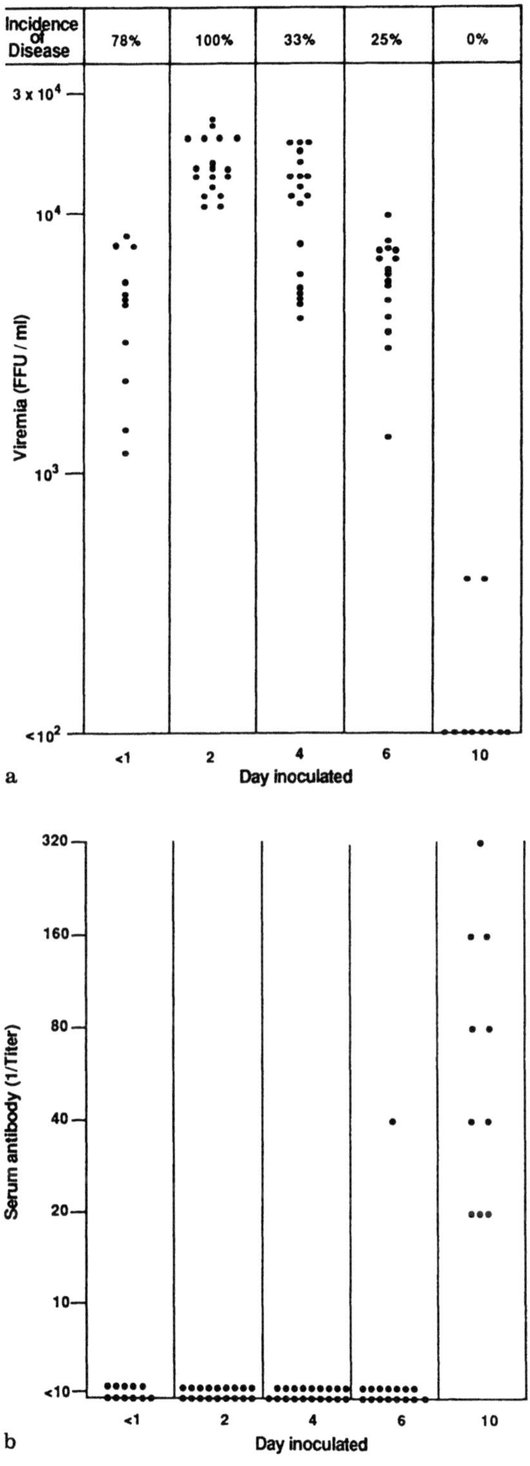

Fig. 2a, b. IRW mice were inoculated intraperitoneally with 7×10^3 FFU WM-E at various times after birth. Mice were evaluated weekly for signs of neurologic disease and were bled at 4, 8, and 12 weeks of age. Infectious virus was titered in the serum by a focal immunofluorescence assay, and serum antiviral antibody was titered by enzyme-linked immunoadsorbent assay (ELISA) (PORTIS et al. 1987). Representative data for viremia (**a**) and antibody (**b**) are shown for the 4-week bleeding. Each *dot* represents a single mouse. The results were similar at 8 and 12 weeks. The cumulative incidence of neurologic disease over a 6-month period of observation is shown at the top of **a**. The kinetics of disease for these groups of mice is represented in Fig. 3

very late in disease of inflammatory infiltrates associated with the CNS lesions and (b) susceptibility to disease is dependent on the induction of immunologic tolerance to the virus. Neurologic disease is seen only in mice exposed to WM-E between < 1 and 6 days of age (see top of Fig. 2a). These mice exhibit persistent high-level viremia (Fig. 2a) and immunologic unresponsiveness to the virus (Fig. 2b), whereas mice inoculated on day 10 remain generally nonviremic (Fig. 2a), produce an immune response to the virus (Fig. 2b), and are fully resistant to disease (top Fig. 2a). The tempo of the disease, manifested by the length of the incubation period, is related to the age at which inoculation occurs, mice inoculated at < 1–2 days of age developing paralytic disease earlier than those infected at 4–6 days of age (Fig. 3). Further acceleration is achieved by inoculation

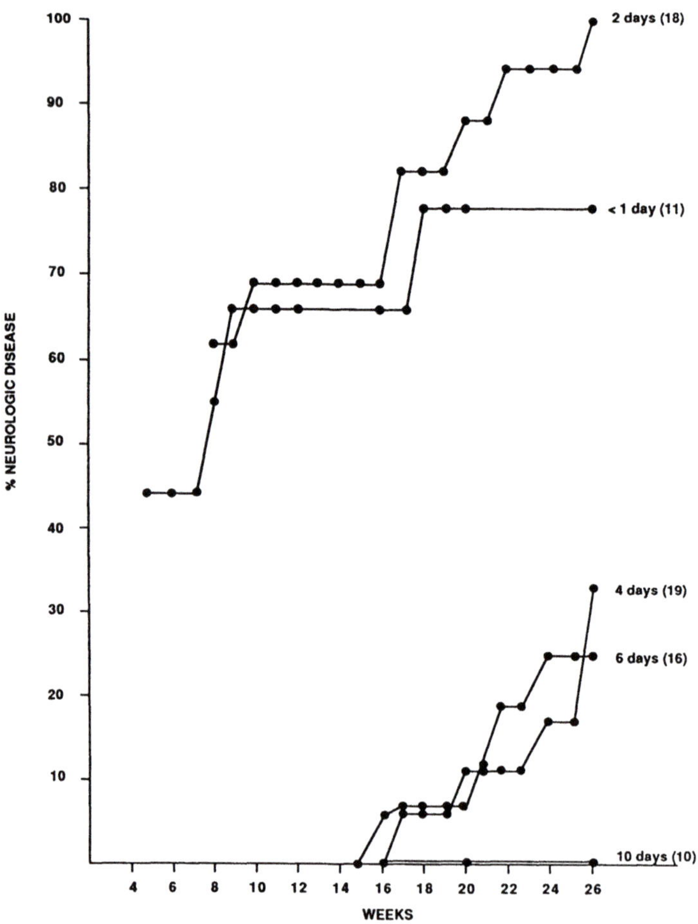

Fig. 3. Kinetics of CNS disease as a function of the time (after birth) of WM-E inoculation. These data were accumulated from the mice shown in Fig. 2a, b. *To the right* of each curve is shown the day after birth that WM-E was inoculated along with the number of mice per group *in parentheses*

in utero (SHARPE et al. 1987). The immune response appears to be the primary mechanism responsible for this age-dependent resistance. The resistance exhibited by mice inoculated at 10 days of age can be overcome by depletion of T-lymphocytes (HOFFMAN et al. 1984), and resistance is conferred to mice inoculated at 2 days of age by adoptive transfer of immune T cells (HOFFMAN et al. 1984). Thus, in contrast to the encephalitides caused by the lentiviruses such as visna, in which the expression of disease is due to indirect damage caused by the host's immune response to the virus (Nathanson et al. 1976), the neuropathology, caused by WM-E appears to be a direct consequence of virus replication in the CNS, the immune response playing a protective role.

2 Virus Transmission

The transmission of this virus has been studies extensively by GARDNER in wild mouse populations (GARDNER et al. 1979). Congenital transmission from mother to offspring is the most important mode of infection in nature. Here I want only to describe a study which we carried out on sexual transmission of the virus (PORTIS et al. 1987). Although WM-E is not unique in its capacity to be transmitted sexually from males to females (sexual transmission of F-MuLV also occurs), transmission by this route is particularly efficient. After a single exposure of superovulated females to viremic males, approximately 50% of the females become infected. This appears due to the unique tropism of WM-E for epithelial cells lining the collecting system of the male reproductive tract. High levels of infectious virus are excreted in the seminal fluid, both cell free and adherent to the surface of spermatozoa.

Females infected by this route develop a brisk immune response which prevents spread of the virus beyond the spleen. These mice do not exhibit a viremia, remain free of neurologic disease, and do not excrete virus. They do not transmit the virus further either to their offspring or other adults. Thus, although this is not an important mode of spread of WM-E within a mouse population, this model should be useful in studying the factors (both viral and host) which promote sexual transmission and may serve as a model for testing the efficacy of vaccine protocols in preventing infection by this route.

3 Qualitative Features of the Virus

WM-E is a typical C-type murine retrovirus, its genome consisting of three structural genes (CHATTOPADHYAY et al. 1981), *gag* (encoding the viral core proteins), *pol* (encoding the viral reverse transcriptase and endonuclease), and *env* (encoding the viral envelop proteins) as well as long-terminal repeats (LTRs) reiterated at each end of the viral DNA which contain the integration signal, and transcriptional initiation, termination, and regulatory sequences. The virus has an ecotropic host range (that is it infects only mouse and rat cells in vitro) and is N-tropic (its replication is restricted in mice of the *FV-1b* genotype) (BRYANT and KLEMENT 1976). Like other retroviruses of the ecotropic host range, WM-E induces syncytia in rat XC cells as well as several other fibroblastic cell lines of mouse origin such as NIH 3T3 and a cell line from *Mus dunni* (PORTIS et al. 1985). Thus, the virus exhibits no obvious features in its pattern of in vitro replication which distinguish it from other murine leukemia viruses. There is no evidence that it is cytopathic for cell lines of neuronal origin including both astroglial and neuronal cell lines which can be chronically infected and maintained for extended periods in culture without the need for repeated addition of uninfected cells (PORTIS, unpublished observations).

Serologic studies using monoclonal antibodies indicate that the viral envelope protein gp70 (surface glycoprotein) exhibits unique epitopes which distinguish WM-E from other exogenous and endogenous murine retroviruses (MCATEE and PORTIS 1985). Two neutralizing epitopes have been identified which are both type-specific. Using intragenic recombinant viruses constructed between Mo-MuLV and WM-E, one of these determinants maps to the N-terminal half and the other to the C-terminal half of the gp70 molecule. Preliminary binding studies indicate that the antibodies defining these structures cross-compete, suggesting their close association in the native molecule.

The viral envelope gene of WM-E has been sequenced (RASSART et al. 1986) and the gp70-coding region exhibits extensive differences at the amino acid level when compared with other murine ecotropic retroviruses. These differences are distributed throughout the gene and indicate that WM-E has evolved independently of other laboratory strains of murine retroviruses. Although the deduced amino acid sequence of the WM-E transmembrane envelope protein (p15E) is highly conserved, this protein nevertheless exhibits at least one epitope which distinguishes it from the transmembrane proteins of other murine retroviruses (MCATEE and PORTIS 1985).

WM-E does exhibit one unique feature which appears to set it apart from other murine retroviruses. In vitro infection of fibroblastic cell lines results in the accumulation of large amounts of unintegrated linear as well as circular DNA (Fig. 4). This unintegrated DNA persists for long periods in clones of NIH 3T3 cells; clone 3A (Fig. 4a) had undergone, since initial infection, approximately 50 cell divisions at the time of analysis. This accumulation of unintegrated DNA appears to be particularly characteristic of the retroviruses which induce cytopathology such as the subgroup B avian leukosis viruses (KESHET and TEMIN 1979), the lentiviruses (SHAW et al. 1984; HARRIS et al. 1984), and the C-type feline retrovirus associated with an immunodeficiency syndrome (MULLINS et al. 1986). At least for the avian viruses and lentiviruses the mechanism appears to involve repeated rounds of superinfection due to a delay in the appearance of viral interference (reviewed by TEMEN 1988). This is probably

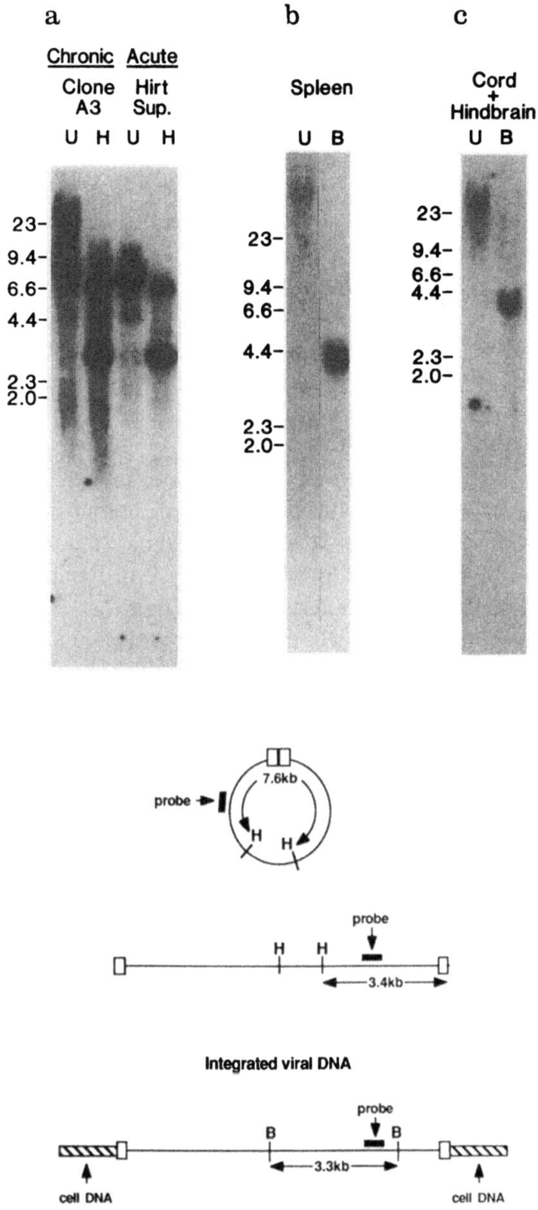

Fig. 4a. Southern bolt showing the persistence of unintegrated circular and linear viral DNA in vitro. NIH 3T3 cells were infected with an MOI of 2 in the presence of 8 μg/ml polybrene, passed three times 1/20, and then cloned by limiting dilution. Clone A3 had undergone approximately 50 cell divisions since initial infection. Total cellular DNA was extracted and cut with *Hin*dIII (*H*), which yields a 7.6-kb fragment derived from unintegrated circular DNA and a 3.4-kb fragment derived from unintegrated linear DNA. Unintegrated 8.7-kb linear DNA can be seen in the uncut lane (*U*), but supercoiled viral DNA is not well visualized. Unintegrated viral DNA derived from a Hirt supernatant of acutely infected cells is included for comparison. **b** and **c** show Southern blots of total tissue DNA extracted from a paralyzed 6-week-old IRW mouse. The DNA was analyzed uncut (*U*) or cut with *Bam*H1 (*B*), which cuts out an internal fragment of 3.3 kb. In the uncut lanes there is no evidence of unintegrated viral DNA; all of the detectable DNA migrated with the high molecular weight cellular DNA. The probe was a 0.5-kb *Xba-Bam*HI fragment encompassing a region of the *env* gene which has been found to be polymorphic in a variety of murine retroviruses (CHAN et al. 1980; BUCKLER et al. 1982; RASSART et al. 1986). Hybridization was carried out at 65°C and blots were washed at the same temperature in 0.1 XSSPE, conditions under which no signal is detected in uninfected NIH 3T3 cells or uninfected mice (not shown)

not the case for WM-E since cell clones, in which unintegrated DNA persists, exhibit high levels of viral interference. In contrast, although low levels of unintegrated viral DNA can be detected in the CNS 3 weeks after infection (not shown), this extrachromosomal viral DNA disappears at later time points when the majority of the viral DNA in the spleen (Fig. 4b) and CNS (Fig. 4c) is integrated.

4 Permissiveness of CNS Vascular Endothelium to Murine Retrovirus Infection

A variety of retroviruses which are not neurovirulent nevertheless replicate in the CNS (OLDSTONE et al. 1983). We found in a survey of viruses of both ecotropic and polytropic host ranges that there was no correlation between the level of CNS replication and neurovirulence (Table 1). When examined by immunohistochemical techniques with monoclonal antibodies specific for the respective viral glycoproteins, two of the nonneurovirulent viruses Akv and F-MuLV (strain FB-29) were expressed exclusively in cells associated with the CNS vasculature (endothelial cells and perithelial cells). At least for F-MuLV, replication appeared to be fully permissive since electron micrographs reveal virus budding from both luminal and basal surfaces of the endothelial cells.

Ultrastructural studies (ANDREWS and GARDNER 1974; SWARZ et al. 1981; PITTS et al. 1987) also indicate that WM-E replicates in cells associated with the CNS vasculature. However, this virus does not remain confined to blood vessels but can be detected in extravascular elements predominantly consisting of neurons (GARDNER et al. 1973; ANDREWS and GARDNER 1974; OLDSTONE et al. 1977) and oligodendroglial cells (OLDSTONE et al. 1977). Using type-specific monoclonal antibodies we have confirmed the extravascular spread of the virus but the precise nature of the cells expressing viral protein has not been resolved. These observations suggest that the neurovirulence of WM-E appears to be associated with its capacity to spread extravascularly within the CNS.

It should be mentioned that the ultrastructural evidence of virus replication within neurons (Fig. 1) has been controversial (ANDREWS and GARDNER 1974; SWARZ et al. 1981; PITTS et al. 1987). The possibility that virus particles found budding within these cells represent the activation of an endogenous virus has not been excluded, but with the availability of WM-E-specific reagents as well as neuronal cell markers this issue should soon be resolved. The problem becomes more troublesome when one considers the fact that proviral integration appears to require actively dividing cells. Replication of a retrovirus in neurons which are

Table 1. Replication of murine retroviruses in the CNS

Virus	Host range	Neurologic disease[a]	Viremia FFU/ml serum[b]	Spleen IC/10^6 cells[b]	Lumbar cord IC/g[c]
WM-E	Ecotropic	+	4×10^4	1×10^4	2×10^5
F-MuLV	Ecotropic	–	6×10^5	3×10^5	1×10^5
98D13	Polytropic	+	1×10^4	2×10^4	3×10^3
F-MCF-1	Polytropic	–	5×10^2	2×10^3	2×10^3

[a] The paralytic disease induced by WM-E and 98D13 are clinically and pathologically similar
[b] Serum virus as well as splenic and lumbar cord infectious centers (IC) were quantitated by a focal immunofluorescence assay using monoclonal antibodies (PORTIS et al. 1985)
[c] Segments of lumbar spinal cord were weighed and dissociated with 0.25% trypsin/collagenase prior to seeding the cells onto indicator cells. The enzyme treatment has been shown to be highly effective at inactivating extracellular virus from contaminating plasma in the tissues (MCATEE and PORTIS 1985)

postmitotic at birth and in which DNA replication is markedly suppressed would suggest a special mechanism for WM-E replication in these cells. The possibility exists that the viral genome might be maintained in neurons in an unintegrated form, and there is some evidence that unintegrated retroviral DNA can serve as a template for viral gene transcription (Harris et al. 1984). However, this seems unlikely in view of the paucity of unintegrated DNA detected in the CNS (Fig. 4c).

5 Genetic Resistance to CNS Disease

Since WM-E is an N-tropic retrovirus, its replication is restricted in *FV-1*^{b} mice (e.g., Balb/c, C57BL) and these mouse strains are resistant to the CNS disease (Oldstone et al. 1980). However, among mice of the *FV-1*^{n} genotype, there is strain variation in disease susceptibility and two mouse strains (AKR and NZB) are fully reistant (Hoffman and Morse 1985). It appears that this disease resistance is due to multiple host genes and is thus difficult to study using standard genetic approaches.

We were interested in determining whether WM-E replication is restricted in these resistant mice. However, since both resistant strains express high levels of endogenous retroviruses, WM-E-specific reagents had to be developed. We prepared monoclonal antibodies by inoculating adult AKR mice with WM-E. Since these mice are tolerant to the antigens of their endogenous virus (Huebner et al. 1971), antiviral antibodies which were made by these mice were predominantly WM-E specific (McAtee and Portis 1985). Monoclonal antibodies derived from these mice reacted with envelope (gp70, p15E) and matrix (p15gag) proteins and a number of antibodies to the envelope proteins were highly specific for WM-E, being nonreactive with endogenous viruses (ecotropic, xenotropic, or polytropic). Using these reagents we found that despite the fact that AKR mice are resistant to the neurologic disease induced by WM-E, the virus replicated in the spleen and CNS at levels equivalent to that of susceptible mouse strains such as NFS and IRW (McAtee and Portis 1985). Preliminary immunofluorescence studies indicate that the replication of WM-E in the CNS of AKR mice is limited to the vasculature, suggesting that the resistance of AKR mice is likely a function of the restriction of extravascular spread of the virus.

6 Latency of Neurological Disease

It is now clear that the length of the latent period is influenced by several factors including mouse strain (Oldstone et al. 1980; Hoffman and Morse 1985), age at which virus is inoculated (Fig. 3), site of inoculation (DesGroseillers et al. 1985), and virus concentration of the inoculum (Brooks et al. 1979). Because of the relatively long latency of the neurologic disease induced by WM-E (up to 1 year in wild mice), this disease has been included among the so-called slow virus diseases (OLDSTONE et al. 1977). The slowness of the clinical disease, however, is not reflected in a slowness of WM-E infection of the CNS. Viral protein (OLDSTONE et al. 1983) as well as viral DNA (PORTIS, unpublished observations) can be detected in the CNS within 1–2 weeks after neonatal inoculation. This suggested that the delay in onset of clinical disease may reflect a requirement for generation of recombinant retroviruses similar to those involved in the induction of certain long-latency leukemias. Although WM-E induces the generation of recombinant polytropic viruses in the spleen (HOFFMAN et al. 1981; OLDSTONE et al. 1983) [which may be involved in the infrequent occurrence of lymphoma (RASSART et al. 1986)], these viruses have not been detected in the CNS (HOFFMAN et al. 1981; OLDSTONE et al. 1983). Thus, the long latency is probably more a function of slowness of spread of WM-E among specific target cells in the CNS. In addition, pathologic studies indicate that, despite the lack of clinical signs of disease, CNS lesions can already be detected within 1–2 weeks after virus inoculation (OLDSTONE et al. 1983). This suggests that the slowness of disease onset may, in part, be a function of the capacity of the CNS to absorb a certain level of damage before clinical signs are manifested.

7 Neurovirulence of WM-E Is Determined by the Viral *env* Gene

The importance of the viral *env* gene in disease pathogenesis was initially suggested by peptide-mapping studies (OLDSTONE et al. 1980) and has subsequently been proven by construction of chimeric viruses. An infectious clone of WM-E was obtained by Jolicoeur from unintegrated circular viral DNA in a Hirt supernatant preparation isolated from NIH 3T3 cells acutely infected with the virus (JOLICOEUR et al. 1983). The virus rescued from NIH 3T3 cells transfected with the cloned DNA, when inoculated into susceptible strains of mice, caused paralytic disease with a latency of 4–5 months. This clone was subsequently used to construct recombinant viruses with amphotropic virus serving as the nonneurovirulent parent (DESGROSEILLERS et al. 1984). This approach revealed that the neurovirulent phenotype of WM-E mapped within a region extending from a *Sal* site near the middle of the *pol* gene to a *Cla* site approximately 95 bases 5′ of the p15E termination codon. In view of the high degree of homology of WM-E and amphotropic virus within the 3′ *pol* sequences (RASSART et al. 1986), it is likely that neurovirulence is a property of the *env* gene. Mapping studies indicate that the neurovirulence of the *ts*1 mutant of Mo-MuLV also maps within the viral *env* gene (SZUREK et al. 1988).

Although it might seem obvious that tissue tropism of a virus is determined by the capacity of the viral surface protein to interact with tissue-specific receptors, such organ- or cell-type-specific receptors have not been identified in any retroviral system other than HIV. Murine and avian retroviral receptors have been identified functionally based primarily on the species specificity of a particular retrovirus and on its capacity to interfere with infection by other retroviruses (interference grouping) (REIN and SCHULTZ 1984). Doubts as to the importance of tissue-specific receptors in determining disease specificity have come primarily from the finding that the disease specificity of the long-latency murine leukemia viruses is determined not by the viral *env* gene but by the viral LTR (DESGROSEILLERS and JOLICOEUR 1984; CHATIS et al. 1984; LI et al. 1987b) (discussed further below). However, these diseases are not induced directly by the input virus, but instead depend on the generation of recombinant polytropic viruses which appear to be the proximal agents responsible for leukemogenesis (RUSCETTI et al. 1981; BULLER et al. 1988). The recombinations result in the acquisition by these viruses of *env* sequences derived from endogenous retroviruses. The importance of these *env* substitutions of the 'ultimate pathogenic viruses' in determining disease specificity is currently not understood.

The probable importance of the envelope gene in determining retroviral disease specificity has come primarily from studies of the acute erythroblastosis induced by the replication-defective component of the Friend complex, SFFV (LINEMEYER et al. 1981; RUTA et al. 1983). Its *env* gene encodes a truncated glycoprotein, gp52, which has been definitively shown to provoke acute erythroproliferation in the absence of other SFFV structural genes (although virus spread requires the presence of helper virus) (WOLFF and RUSCETTI 1988). The mechanism is currently thought to involve a receptor for gp52 (LI et al. 1987a) but its nature, or whether it serves a normal function in erythroid differentiation, is unknown.

The function of the viral *env* gene in the pathogenesis of the paralytic disease is being studied using chimeric viruses constructed between F-MuLV strain FB-29 (SITBON et al. 1986), which induces long-latency (2–3 months) erythroleukemia, and a noninfectious molecular clone of WM-E strain 1504E (p12-4), cloned from a Hirt supernatant of acutely infected NIH3T3 cells. Although the p12-4 genome is replication defective, the defect is located outside the *env* gene, which functions normally if introduced into a replication-competent virus. The chimeric virus 10A3 contains the LTR *gag* and *5' pol* genes from FB-29 and *3' pol* and *env* genes from p12-4 (Fig. 5). The virus is infectious for mice and induces both a neurologic disease with a latency of approximately 10–12 weeks as well as a long-latency myeloid (mixed erythroid/granulocytic) leukemia with a latency of 5–6 months (Fig. 5). Thus, the chimeric virus has the disease specificities of both parent retroviruses with two exceptions. The latency of the leukemia induced by 10A3 was longer than that caused by FB-29 and it was of mixed cell type, whereas FB-29 causes a pure erythroleukemia (SITBON et al. 1986). As expected, by immunofluorescence the distribution of 10A3 within the CNS was identical to that seen with WM-E (i.e., both vascular and extravascular) whereas, as mentioned above, the FB-29 parent replicated only in the blood vessels of the CNS. This suggests that the *pol-env* sequence of WM-E determines the capacity of the virus to spread extravascularly.

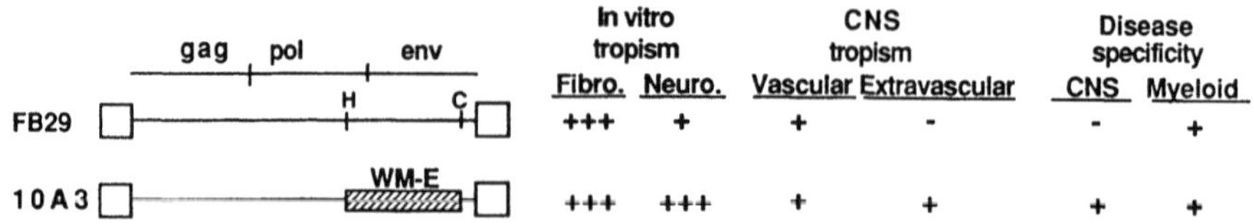

Fig. 5. Schematic summary of the tropism and disease specificity of Friend MuLV (strain FB-29) and a chimeric virus 10A3 constructed of the FB-29 genome but containing a *Hind* III-Cla fragment from a noninfectious molecular clone of WM-E strain 1504E (p12-4)

Although the *env* gene of WM-E endows the virus with the capacity to spread within the CNS, this does not establish whether the *env* protein actually recognizes unique receptors on neuronal cells. This question can best be studied in vitro. Unfortunately, mouse neuroblastoma cell lines characteristically express endogenous retroviral gp70 which has been serologically typed as ecotropic and thus might be expected to interfere with infection by other ecotropic viruses. Nevertheless, the infectivity of FB-29 and the chimeric virus 10A3 were compared using two neuroblastoma lines C1300 and Clone NA both of which are clones of Neuro 2A and express low but detectable levels of endogenous gp70. Although these cell lines were originally derived from the FV-1b mouse strain A/J, both FB-29 and 10A3 are n/b tropic and thus infect these cells with one-hit kinetics. The titer of the viruses on NIH3T3 cells (FB-29: 8×10^5 focus-forming units (FFU)/ml; 10A3: 2×10^5/ml) was used to calculate 'FFU added' on the horizontal

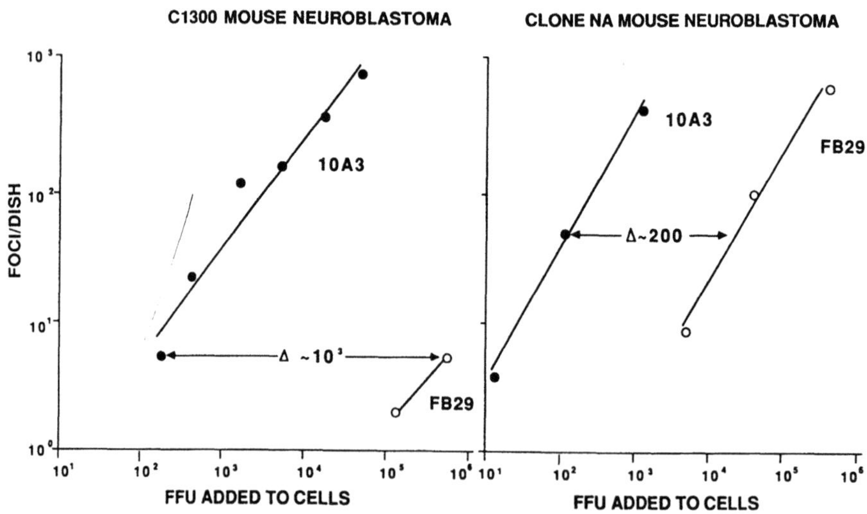

Fig. 6. Comparison of the infectivity of FB-29 and the chimeric virus 10A3 for two neuroblastoma cell lines. $1-2 \times 10^5$ cells seeded into 60-mm dishes were infected with dilutions of virus in the presence of 8 g/ml Polybrene. Foci were scored with a focal immunofluorescence assay using monoclonal antibodies specific for F-MuLV gp70 (FB-29) and WM-E gp70 (10A3)

axes in Fig. 6. By virtue of the use of monoclonal antibodies specific for the infecting virus, this assay detects foci of infection only of the added virus and does not detect the endogenous gp70. This approach clearly showed that 10A3 exhibited enhanced infectivity for neuroblastoma cells, relative to its parent virus FB-29 even in the presence of endogenous ecotropic retroviral *env* protein (Fig. 6). Since these viruses differ by only their *env* genes and *3' pol* sequences, these observations support the notion that these neuronal cell lines express receptors which preferentially bind the *env* glycoproteins of WM-E. Binding studies with purified glycoproteins need to be done to test this hypothesis.

8 Influence of Viral LTR on Neurovirulence

In discussing the function of the retroviral LTR, one has to distinguish between its role in controlling virus replication from its role in determining disease specificity. It is clear that among the MuLVs which induce long-latency hematopoietic neoplasms, enhancer sequences within the U3 region of the LTR have a profound influence on the cell type which is ultimately transformed (DesGroseillers and Jolicoeur 1984; Chatis et al. 1984; Li et al. 1987b). Yet the tissue specificity exhibited for different LTRs in transient transcription assays does not always correlate with the disease specificity encoded by the respective LTRs. Thus, although Moloney MuLV induces thymomas but not erythroleukemias, its LTR functions well in both thymocytes and erythroblasts (Spiro et al. 1988). In addition, F-MuLV, which harbors a strongly erythrotropic enhancer (Short et al. 1987), induces granulocytic leukemias and lymphomas in certain mouse strains (Silver and Fredrickson 1983; Chesebro et al. 1983). These discrepancies serve to reinforce the complexities of these systems. Induction of long-latency hematopoietic neoplasms is, at least in the case of Mo-MuLV-induced thymoma and F-MuLV-induced erythroleukemia, a multistep process which involves the generation of recombinant viral intermediates, cellular proliferative stimuli, and activation of specific cellular *onc* genes. Which of these steps is influenced by the LTR of the input virus is unknown at this time.

In the case of the neurologic disease caused by WM-E, neuropathology is tightly linked to the replication of the input virus within the CNS and appears not to involve other viral intermediates. LTRs from a number of murine retroviruses, which have distinct hematopoietic disease specificities, nevertheless function within the CNS (Fig. 5) (DesGroseillers and Jolicoeur 1984; DesGroseillers et al. 1985). Yet the hematopoietic disease specificities of the respective LTRs are retained in the chimeric viruses. Thus, WM-E containing an LTR from Mo-MuLV also causes thymomas in a majority of inoculated recipients (DesGroseillers et al. 1985). Similarly, 10A3 containing the FB-29 LTR causes myeloid leukemias (Fig. 5). These observations further distinguish the effect of the LTR in determining hematopoietic disease specificity from its effect on virus replication.

There is evidence suggesting that the LTR may have subtle effects on qualitative features of the CNS disease. A chimeric virus constructed of the WM-E genome but containing the LTR and 3′ p15E sequences from Mo-MuLV was found to induce a neurologic disease somewhat different than that caused by the parent WM-E (DesGroseillers et al. 1985). Lesions were seen in the cerebral cortex and striatum, regions which are not affected by WM-E. Although virus localization studies were not carried out, this observation suggests that sequences within the viral LTR may influence the sites of virus replication within the CNS.

9 Concluding Remarks

The nature of the neurologic diseases caused by the murine retroviruses is complex. Their pathogenesis appears not to involve immunopathology, which sets these diseases apart from those caused by lentiviruses such as visna and caprine arthritis encephalitis virus. The wild mouse disease appears to be a direct consequence of the replication of the virus in the CNS, a function which is determined primarily by viral envelope gene. Whether the envelope proteins function to target the viral genome to the right cells or whether the WM-E *env* proteins themselves are toxic in the CNS is unknown.

WM-E does not replicate to very high levels in the CNS. In fact the level of infectious centers in the spleen is on the order of 100-fold higher than the CNS (McAtee and Portis 1985) when virus assays are standardized with respect to cellular DNA content of the respective organs. In addition, the finding of vacuolar degeneration in the CNS, in the absence of recognizable virions, and the consistent observation of aberrant virus assembly in neurons and oligodendrocytes, suggest that replication of WM-E within the CNS may be defective. In support of this hypothesis, Oldstone found that WM-E replication in vivo is associated with the accumulation of viral p30 core protein relative to viral *env* protein gp70 (Oldstone et al. 1980). Although reminiscent of the *env* protein-processing defect of the neurovirulent *ts*1 mutant of Mo-MuLV (Wong et al. 1983), defective cleavage has not been observed after in vitro infection with WM-E (McAtee and Portis 1985).

It is intriguing to speculate that one or both of the viral envelope proteins may be toxic when produced within the CNS. A constant source of envelope protein might be provided by the endothelial and perithelial cells which appear to be highly permissive for retrovirus replication. An acute vacuolar myelopathy induced by a *ts* mutant of VSV (Dal Canto et al. 1976) (the replication of which is restricted in the CNS) was found to be associated with the accumulation of G protein (the viral *env* protein) within the vacuolar lesions (Robain et al. 1986). The fusigenic properties of this protein in vitro have suggested that its accumulation in the CNS might perturb membrane function either intra- or extracellularly. Membrane fusion requires an intimate association between the fusigen and the membrane, a function usually fulfilled by a receptor for the viral glycoprotein. Thus, studies of this aspect of pathogenesis of the wild mouse neurologic disease will likely require learning more about the nature of the viral receptor as well as the regulation of gene expression within the CNS at both the RNA and protein level. Further construction of chimeric viruses, perhaps focussing on parental viruses of common ancestry but of differing virulence, promises to define the *env* sequences which determine neurotropism as well as the role of transcriptional regulatory sequences within the LTR in influencing the sites of viral expression within the CNS. These efforts, in addition to answering questions related to pathogenesis of retroviral diseases of the nervous system, should also contribute to our understanding of the factors which may be useful in future attempts to target exogenous genes to specific sites within the CNS.

Acknowledgments. I wish to thank Stanley Hayes for electron microscopy, Frank McAtee for technical assistance, and Irene Cook Rodreguez for manuscript preparation.

References

Andrews JM, Gardner MB (1974) Lower motor neuron degeneration associated with type C RNA virus infection in mice: neuropathological features. J Neuropathol Exp Neurol 33: 285–307

Brooks BR, Swarz JR, Narayan O, Johnson RT (1979) Murine neurotropic retrovirus spongiform polioencephalomyelopathy: acceleration of disease by virus inoculum concentration. Infect Immun 23: 540–544

Brooks BR, Swarz JR, Johnson RT (1980) Spongiform polioencephalomyelopathy caused by a murine retrovirus. Lab Invest 43: 480–486

Bryant ML, Klement V (1976) Clonal heterogeneity of wild mouse leukemia viruses: host range and antigenicity. Virology 73: 532–536

Buckler CE, Hoggan MD, Chan HW, Sears JF, Khan AS, Moore JL, Hartley JW, Rowe WP, Martin MA (1982) Cloning and characterization of an envelope-specific probe from xenotropic murine leukemia proviral DNA. J Virol 41: 228–236

Buller RS, Sitbon M, Portis JL (1988) The endogenous MCF gp70 linked to the *Rmcf* gene restricts MCF virus replication *in vivo* and provides partial resistance to erythroleukemia induced by Friend murine leukemia virus. J Exp Med 167: 1535–1546

Chan HW, Bryan T, Moore JL, Staal SP, Rowe WP, Martin MA (1980) Identification ecotropic proviral sequences in inbred mouse strains with a cloned subgenomic DNA fragment. Proc Natl Acad Sci USA 77: 5779–5783

Chatis PA, Holland CA, Silver JE, Frederickson TN, Hopkins N, Hartley JW (1984) A 3′ end fragment encompassing the transcriptional enhancers of nondefective Friend virus confers erythroleukemogenicity on Moloney leukemia virus. J Virol 52: 248–254

Chattopadhyay SK, Oliff AI, Linemeyer DL, Lander MR, Lowy DR (1981) Genomes of murine leukemia viruses isolated from wild mice. J Virol 39: 777–791

Chesebro B, Portis JL, Wehrly K, Nishio J (1983) Effect of murine host genotype of MCF virus expression, latency, and leukemia cell type of leukemias induced by Friend murine leukemia helper virus. Virology 128: 221–233

Dal Canto MC, Rabinowitz SG, Johnson TC (1976) An ultrastructural study of central nervous system disease produced by wild-type and temperature-sensitive mutants of vesicular stomatitis virus. Lab Invest 35: 185–196

DesGroseillers L, Jolicoeur P (1984) Mapping the viral sequences conferring leukemogenicity and disease specificity in Moloney and amphotropic murine leukemia viruses. J Virol 52: 448–456

DesGroseillers L, Barrette M, Jolicoeur P (1984) Physical mapping of the paralysis-inducing determinant of a wild mouse ecotropic neurotropic virus. J Virol 52: 356–363
DesGroseillers L, Rassart E, Robitaille Y, Jolicoeur P (1985) Retrovirus-induced spongiform encephalopathy: The 3'-end long terminal repeat-containing viral sequences influence the incidence of disease and the specificity of the neurological syndrome. Proc Natl Acad Sci USA 82: 8818–8822
Gardner MB, Chivi A, Dougherty MF, Casagrande J, Estes JD (1979) Congenital transmission of murine leukemia virus from wild mice prone to development of lymphoma and paralysis. J Natl Cancer Inst 62: 63–69
Gardner MB, Henderson BE, Officer JE, Rongey RW, Parker JC, Oliver C, Estes JD, Huebner RJ (1973) A spontaneous lower motor neuron disease apparently caused by indigenous type-C RNA virus in wild mice. J Natl Cancer Inst 51: 1243–1254
Harris JD, Blum H, Scott J, Traynor B, Ventura P, Haase A (1984) Slow virus visna: reproduction *in vitro* of virus from extrachromosomal DNA. Proc Natl Acad Sci USA 81: 7212–7215
Hoffman PM, Morse HC (1985) Host genetic determinants of neurological disease induced by Cas-Br-M murine leukemia virus. J Virol 53: 40–43
Hoffman PM, Davidson WF, Ruscetti SK, Chused TM, Morse HC (1981) Wild mouse ecotropic murine leukemia virus infection of inbred mice: dual-tropic virus expression precedes the onset of paralysis and lymphoma. J Virol 39: 597–602
Hoffman PM, Robbins DS, Morse HC (1984) Role of immunity in age-related resistance to paralysis after murine leukemia virus infection. J Virol 52: 734–738
Huebner RJ, Sarma PS, Kelloff GJ, Gilden RV, Meier H, Myers DD, Peters RL (1971) Immunological tolerance to RNA tumor virus genome expressions: significance of tolerance and prenatal expressions in embryogenesis and tumorigenesis. Ann NY Acad Sci 181: 246–271
Jolicoeur P, Nicolaiew N, DesGroseillers L, Rassart E (1983) Molecular cloning of infectious viral DNA from ecotropic neurotropic wild mouse retrovirus. J Virol 45: 1159–1163
Keshet E, Temin HM (1979) Cell killing by spleen necrosis virus is correlated with a transient accumulation of spleen necrosis virus DNA. J Virol 31: 376–388
Li J-P, Bestwick RK, Spiro C, Kabat D (1987a) The membrane glycoprotein of Friend spleen focus-forming virus: evidence that the cell surface component is required for pathogenesis and that it bonds to a receptor. J Virol 61: 2782–2792
Li Y, Golemis E, Hartley JW, Hopkins N (1987b) Disease specificity of nondefective Friend and Moloney murine leukemia viruses is controlled by a small number of nucleotides. J Virol 61: 693–700
Linemeyer DL, Ruscetti SK, Scolnick EM, Evans LH, Duesberg PH (1981) Biological activity of the spleen focus-forming virus is encoded by a molecularly cloned subgenomic fragment of spleen focus-forming virus DNA. Proc Natl Acad Sci USA 78: 1401–1405
McAtee FJ, Portis JL (1985) Monoclonal antibodies specific for wild mouse neurotropic retrovirus: detection of comparable levels of virus replication in mouse strains susceptible and resistant to paralytic disease. J Virol 56: 1018–1022
Mullins JI, Chen CS, Hoover EA (1986) Disease-specific and tissue-specific production of unintegrated feline leukaemia virus variant DNA in feline AIDS. Nature 319: 333–336
Nathanson N, Panitch H, Palsson PA, Petursson G, Georgsson G (1976) Pathogenesis of visna. II. Effect of immunosuppression upon early central nervous system lesions. Lab Invest 35: 444
Oldstone MBA, Lampert PW, Lee S, Dixon FJ (1977) Pathogenesis of the slow disease of the central nervous system associated with WM 1504 E virus. Am J Pathol 88: 193–212
Oldstone MBA, Jensen F, Dixon FJ, Lampert PW (1980) Pathogenesis of the slow disease of the central nervous system associated with wild mouse virus. II. Role of virus and host gene products. Virology 107: 180–193
Oldstone MBA, Jensen F, Elder J, Dixon FJ, Lampert PW (1983) Pathogenesis of the slow disease of the central nervous system associated with wild mouse virus. III. Role of input virus and MCF recombinants in disease. Virology 128: 154–165
Pitts OM, Powers JM, Bilello JA, Hoffman PM (1987) Ultrastructural changes associated with retroviral replication in central nervous system capillary endothelial cells. Lab Invest 56: 401–409
Portis JL, McAtee FJ, Evans LH (1985) Infectious entry of murine retroviruses into mouse cells. Evidence of a post-adsorption step inhibited by acidic pH. J Virol 55: 806–812
Portis JL, McAtee FJ, Hayes SF (1987) Horizontal transmission of murine retroviruses. J Virol 61: 1037–1044
Rassart E, Nelbach L, Jolicoeur P (1986) Cas BrE murine leukemia virus: sequencing of the

paralytogenic region of its genome and derivation of specific probes to study its origin and the structure of its recombinant genomes in leukemic tissues. J Virol 60: 910–919

Rein A, Schultz A (1984) Different recombinant murine leukemia viruses use different cell surface receptors. Virology 136: 144–152

Robain O, Chany-Fournier F, Cerutti I, Mazlo M, Chany C (1986) Role of VSV G antigen in the development of experimental spongiform encephalopathy in mice. Acta Neuropathol (Berl) 70: 220–226

Ruscetti S, Davis L, Field J, Oliff A (1981) Friend murine leukemia virus-induced leukemia is associated with the formation of mink cell focus-inducing viruses and is blocked in mice expressing mink cell focus-inducing xenotropic viral envelope genes. J Exp Med 154: 907

Ruta M, Bestwick R, Machida C, Kabat D (1983) Loss of leukemogenicity caused by mutations in the membrane glycoprotein structural gene of Friend spleen focus-forming virus. Proc Natl Acad Sci USA 80: 4704–4708

Sharpe AH, Jaenisch R, Ruprecht RM (1987) Retroviruses and mouse embryos: a rapid model for neurovirulence and transplacental antiviral therapy. Science 236: 1671–1674

Shaw Gm, Hahn BH, Arya SK, Groopman JE, Gallo RC, Wong-Staal F (1984) Molecular characterization of human T-cell leukemia (lymphotropic) virus type III in the acquired immune deficiency syndrome. Science 226: 1165–1171

Short MK, Okenquist SA, Lenz J (1987) Correlation of leukemogenic potential of murine retroviruses with transcriptional tissue preference of the viral long terminal repeats. J Virol 61: 1067–1072

Silver JE, Fredrickson TN (1983) A new gene that controls the type of leukemia induced by Friend murine leukemia virus. J Exp Med 158: 493–505

Sitbon M, Sola B, Evans L, Nishio J, Hayes SF, Nathanson K, Garon CF, Chesebro B (1986) Hemolytic anemia and erythroleukemia, two distinct pathogenic effects of Friend MuLV: mapping of the effects to different regions of the viral genome. Cell 47: 851–859

Spiro C, Li J-P, Bestwick RK, Kabat D (1988) An enhancer sequence instability that diversifies the cell repertoire for expression of a murine leukemia virus. Virology 164: 350–361

Swarz JR, Brooks BR, Johnson RT (1981) Spongiform polioencephalomyelopathy caused by a murine retrovirus. II. Ultrastructural localization of virus replication and spongiform changes in the central nervous system. Neuropathol Appl Neurobiol 7: 365–380

Szurek PF, Yuen PH, Jerzy R, Wong PKY (1988) Identification of point mutations in the envelope gene of Moloney murine leukemia virus TB temperature-sensitive paralysis to genic mutant *ts*1: molecular determinants for neurovirulence. J Virol 62: 357–360

Temin HM (1988) Mechanisms of cell killing/cytopathic effects by nonhuman retroviruses. Rev Infect Dis 10: 399–405

Wolff L, Ruscetti S (1988) The spleen focus-forming virus (*SFFV*) envelope gene, when introduced into mice in the absence of other *SFFV* genes, induces acute erythroleukemia. J Virol 62: 2158–2163

Wong PKY, Soong MM, MacLeod R, Gallick GE, Yuen PH (1983) A group of temperature-sensitive mutants of Moloney leukemia virus which is defective in cleavage of *env* precursor polypeptide in infected cells also induces hind-limb paralysis in newborn CFW/D mice. Virology 125: 513–518

Zachary JF, Knupp CJ, Wong PKY (1986) Noninflammatory spongiform polioencephalomyelopathy caused by a neurotropic temperature-sensitive mutant of Moloney murine leukemia virus TB. Am J Pathol 124: 457–468

Moloney Murine Leukemia Virus Temperature-Sensitive Mutants: A Model for Retrovirus-Induced Neurologic Disorders

P. K. Y. Wong

1 Introduction

In the early 1960s Stansly observed in BALB/c mice a neurogenic paralysis of the hind limb associated with the cell-free transmission of a reticulum cell sarcoma, and suggested that the paralytogenic agent and the neoplastic virus could be the same (Stansly 1965). About a decade later, the discovery by Gardner and coworkers (1973) that an isolate of murine leukemia virus (MuLV) of wild mouse origin not only possesses the ability to cause lymphomas but also the ability to induce a nononcogenic, yet fatal disease of the CNS further suggested that MuLV is not only lymphomagenic but also neurovirulent. This naturally occurring MuLV-induced neurologic disorder was later found to be readily and reproducibly transmitted to susceptible laboratory mice (Officer et al. 1973; Oldstone et al. 1977), establishing the fact that this MuLV-related neurologic disease not only occurs in wild mice but can also be induced in laboratory mice (for review see

University of Texas, MD Anderson Cancer Center, Science Park-Research Division, PO Box 389, Smithville, Texas 78957, USA

GARDNER 1985). Since then at least three instances of neurologic disorders induced by MuLV in laboratory mice and rats have been reported. In two of these instances, temperature-sensitive (*ts*) mutants of Moloney murine leukemia virus (MoMuLV) were involved. In the first case a group of *ts* mutants of MoMuLV-TB, a variant of MoMuLV, was found to be able to induce a rapidly progressive paralytic disease in mice (MCCARTER et al. 1977; WONG et al. 1983). The second case, which was reported by BILLELLO and coworkers (1986), also involved a *ts* mutant of MoMuLV designated ts Mo BA-I MuLV, which was initially obtained from Peter Nobis of the University of Hamburg, Federal Republic of Germany. *Ts* Mo BA-I MuLV induced a neurodegenerative disorder characterized by tremor and hind limb weakness which rarely leads to paralysis. The third case involved a Friend strain of MuLV (F-MuLV) which had been passaged several times in rats (KAI and FURATA 1984). This rat- passaged F-MuLV was also shown to be able to induce hind limb paralysis similar to those induced by the wild mouse virus and by the paralytogenic *ts* mutants of MoMuLV-TB. Although the latency and incidence of development as well as severity of the neurologic disorders induced by these different neurovirulent MuLVs are variable, they all share a more or less similar symptomatology and a common histopathologic feature, i.e., noninflammatory spongiform encephalomyelopathy.

Another common cause of hind limb paralysis in MuLV-infected laboratory mice which is different from those mentioned above is an infiltration of leukemic cells into the CNS. This perineural leukemic cell infiltration accounts for the hind limb paralysis observed in some high-MuLV-expressor AKR and BXH-2 recombinant inbred mice (BEDIGIAN et al. 1984).

The spectrum of retroviruses capable of inducing neurodegenerative diseases, however, is not limited to murine leukemia viruses. For example, visna virus, which belongs to a subfamily of retroviruses called lentiviruses, causes CNS damage in goats and sheep. However, visna virus may not be directly responsible for damaging the nerve cells since host inflammatory response is observed in the impaired CNS (for review see HAASE 1986). More importantly, the human immunodeficiency virus (HIV), the etiologic agent of acquired immunodeficiency syndrome (AIDS), has been shown not only to get into the CNS of AIDS patients, but also to attack specific regions and cell types resulting in vacuolar myelopathy and dementia (ANDERS et al. 1986; NAVIA et al. 1986; PETITO et al. 1985; PAHWA et al. 1987). More recently, human T-cell lymphotropic virus (HTLV-1), which is associated with adult T-cell leukemia, has been implicated as the causative agent of tropical spastic paraparesis (TSP), a common paralytic disease in the tropics and Japan (OSAME et al. 1986; ROMAN 1987). The primary clinical feature of the disease is the development of a progressive weakness in the legs which deteriorates to paraparesis. These findings clearly establish the fact that certain retroviruses, in addition to being oncogenic, are also neurovirulent. This raises the possibility that retroviruses may also be involved in other human neurologic disorders which are as yet without known etiologic agents such as multiple sclerosis, amyotrophic lateral sclerosis, Creutzfelt-Jakob disease and Alzheimer's disease.

Whether there is any common mechanism underlying some of these chronic degenerative neurologic disorders is not known. Animal models with clinical and histopathologic patterns similar to the human neurologic diseases are needed for studies to obtain a clearer understanding of the mechanism(s) that generates these disorders. The MoMuLV *ts* mutant-induced neurologic disease in mice provides a safe and convenient animal model for investigating degenerative neurologic disorders which are similar to those observed in man. This animal model also allows a systematic study of the entire disease process including events associated with the early stages of the disease development. Similar studies in the early stages of the disease development in man are extremely difficult if not impossible to define. In addition, since the paralytogenic mutants were derived from a nonparalytogenic parent, the genomic change(s) in the mutants which confers on the virus the paralytogenic potential can be identified. Studies with this animal model should shed light on the molecular mechanisms associated with neurogenic disorders caused by retroviruses.

This review focuses on the principal features of the disorders induced by the paralytogenic *ts* mutants of MoMuLV-TB, which are summarized in Table 1. Specific emphasis will be given on those features which are particularly relevant to understanding the molecular basis for viral neurotropism and neurovirulence. Comparison will be made between the paralytogenic *ts* mutants of MoMuLV-TB and *ts* Mo BA-1 MuLV as well as the neurotropic wild mouse ecotropic viruses (collectively termed WM-E virus in this review) and the rat-passaged F-MuLV model.

Table 1. Principal features of the neurologic disorders induced by *ts*1 mutant of MoMuLV-TB

Host: Susceptible–BALB/c, CFW/D, C3H, CBA, FVB, SWR/J, NIH Swiss. Resistant–C57BL, B10.D2, SENCAR and BALB/c, (nu/nu)

Incidence and latency period: 82%–100% of susceptible mice within 4–12 weeks if inoculated during the first 48 h after birth; dose and age dependent; mice > 10 days of age are generally resistant

Transmission: By intraperitoneal or intracranial inoculation; by perinatal transfer from mother to neonate

Virus properties: Temperature sensitive (39°C restrictive temperature); defective in intracellular processing of envelope precursor polyprotein; NB-tropic, neurotropic, T-cell tropic

Virus pathogenesis: Virus replicates in lymphoid tissues, in particular, thymus and spleen, in the early stages of infection and spread by viremia or via macrophage to other organs including the CNS

Nonspecific polyclonal humoral response to virus

Generation of recombinant viruses with endogenous virus not involved

Symptoms: Growth retardation, scruffy coat, hind limb tremors, progressive symmetric paresis, atrophy of hind limb musculature, hind limb paralysis, severe atrophy of thymus and spleen, secondary infections, immunosuppression

Histopathology:

CNS: Vacuolar degeneration of neural cell bodies and processes in midbrain, brain stem, and gray and white matter of the spinal cord

Swollen and vacuolated glial cells; spongiform changes in associated neuropil

Absence of host inflammatory response

Virions detected by EM in extracellular spaces and in very low number budding from cell membrane of neural cells. Budding viral particles from endothial cells were also observed

Thymus: Severe reduction in size and thinning of cortex. Many pyknotic nuclei within the infected thymic cortex

2 Characteristics of the Paralytogenic Temperature-Sensitive Mutants

Moloney murine leukemia virus was originally isolated in complex with a sarcoma virus by MOLONEY (1960) from the transplantable mouse sarcoma 37. MoMuLV is a leukemia virus which normally induces a T-cell lymphoma in 100% of the neonatally infected mice within 3–6 months postinoculation (p.i.) (MOLONEY 1960; REDDY et al. 1980). MoMuLV-TB is a variant of MoMuLV (WONG et al. 1973). MoMuLV-TB induces T-cell lymphoma in about 30% of infected mice in 4–10 months p.i. (YUEN and SZUREK 1989).

The MoMuLV genome, like most of the retroviruses which do not carry an oncogene, consists of three genes: *gag*, *pol*, and *env* (Fig. 1). The *gag* gene encodes for a polyprotein precursor, $Pr65^{gag}$, which is later cleaved by a viral encoded protease into four structural proteins which make up the core of the virion. Read-through of the termination signal of the gag gene results in the translation of a larger precursor, $Pr180^{gag\text{-}pol}$. $Pr180^{gag\text{-}pol}$ is posttranslationally cleaved into three viral proteins: the *gag* protease, the reverse transcriptase, and an integrase which is necessary for integration. The *env* gene encodes for a large precursor polyprotein $Pr80^{env}$ which is subsequently cleaved by a cellular protease at two sites to yield gp70 and Prp15E. The amino-terminal 33 amino acids constitute a

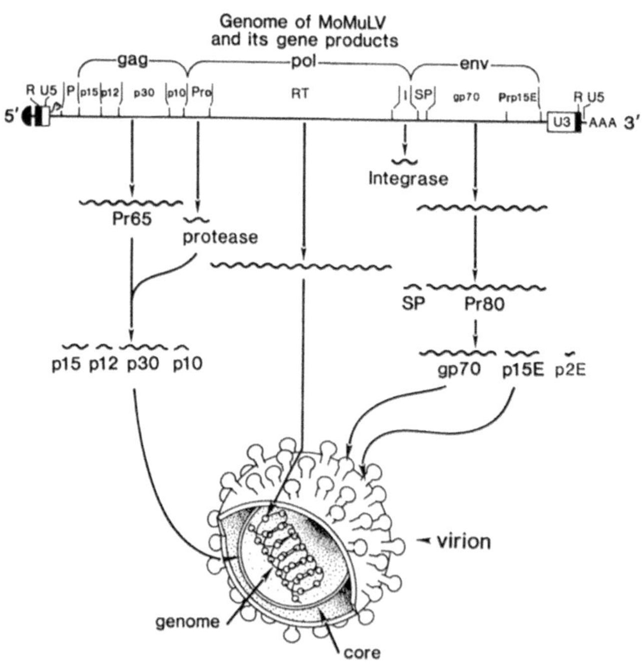

Fig. 1. Schematic presentation of MoMuLV genome, its gene products and virus particle. Packaging sequence; *Pro* protease; *RT* reverse transcriptase; *I* integrase; *SP* signal peptide

signal peptide which is cleaved in the rough endoplasmic reticulum (RER). The gp70 and Prp15E cleavage products are found in the *cis* compartment of the Golgi complex. Prp15E is processed to p15E and p2E by the viral-encoded protease. The large glycoprotein, gp70, forms the surface "knobs" or "spikes" and contains the host cell receptor-binding site as well as the major sites recognized by neutralizing antibodies. The p15E protein can be divided into three functional regions, an external region which anchors with gp70 to the virion, a highly hydrophobic transmembrane region, and a carboxy-terminal cytoplasmic domain (for review see GOFF and LOBEL 1987).

Both spontaneous and bromodeoxiuridine-induced temperature-sensitive mutants have been isolated from MoMuLV-TB (RUDE et al. 1980; WONG and GALLICK 1978; WONG et al. 1973). A group of such *ts* mutants, *ts*1, *ts*7, and *ts*11, was shown to be capable of inducing the neurogenic hind limb paralysis on inoculation into neonates of susceptible mice (MCCARTER et al. 1977; WONG et al. 1983). It is interesting to note that these three paralytogenic mutants were isolated on separate occasions and that, although *ts*1 and *ts*7 are spontaneous mutants (WONG et al. 1973), *ts*11 was isolated after treatment of MoMuLV-TB with bromodeoxyuridine (RUDE et al. 1980). These findings establish that the generation of the paralytogenic phenotype was not an isolated occurrence. Temperature sensitivity per se, however, is not necessarily an essential property for neurologic disease induction. Other *ts* mutants of MoMuLV-TB, such as *ts*3, fail to induce paralysis in mice which are susceptible to the paralytogenic *ts* mutants (MCCARTER et al. 1977; WONG et al. 1983). Another *ts* mutant of MoMuLV, *ts* Mo BA-1 clone 14, which has marked temperature sensitivity, has been shown to be much less effective in inducing neurologic disease than uncloned *ts* Mo BA-1 MuLV (BILELLO et al. 1986). Furthermore, GREENBERGER et al. (1975) reported that newborn mice inoculated with *ts* mutant isolates of Rauscher murine leukemia virus showed no paralytic disease.

Biochemical studies of the *ts* mutant of MoMuLV-TB showed that in addition to the temperature-sensitive phenotype, i.e., reduced infectious virus production at the restrictive temperature (39°C), the paralytogenic *ts* mutants, *ts*1, *ts*7, and *ts*11, are all defective in the processing of the envelope precursor polyprotein, $Pr80^{env}$, into gp70 and Prp15E at the restrictive temperature (WONG et al. 1983). Since the nonprocessed envelope protein accumulates intracellularly and does not incorporate into the virion, virus particles are produced from the infected cells with reduced amounts of envelope proteins when compared with the parental MoMuLV-TB. Consequently, the uncleaved envelope precursor polyprotein accumulates in the infected cell. In contrast, the nonparalytogenic mutant, *ts*3, which is defective in assembly and release of virus particles (WONG and MCCARTER 1974; YUEN and WONG 1977), shows normal processing of $Pr80^{env}$ in the infected cells at 39°C (WONG et al. 1983). Since the paralytogenic mutants are defective in processing of the $Pr80^{env}$ while the nonparalytogenic *ts*3 mutant and the parental MoMuL-TB process $Pr80^{env}$ normally, it suggests that the paralytogenic ability of *ts*1, *ts*7, and *ts*11 may be due to alteration(s) of the viral genome and that the nonprocessing of $Pr80^{env}$ may be linked to the neurologic pathogenesis. This hypothesis is further substantiated by molecular genetic studies (Wong et al. 1985; Yuen et al. 1985 a, b, 1986; Szurek et al. 1988), which will be discussed in a later section of this review.

While the intracellular accumulation of the uncleaved $Pr80^{env}$ precursor may be linked to paralysis induction in studies involving the *ts*1, *ts*7, and *ts*11mutants, this may represent only one of the mechanisms of pathogenesis involved in the neurologic diseases induced by mutants of MoMuLV. Studies by Bilello and coworker (1986) indicate that *ts* Mo BA-1 MuLV, which induces a slower progressive neurodegenerative disease than *ts*1, *ts*7, and *ts*11, is temperature sensitive in a late function and processes $Pr80^{env}$ normally at both the permissive and the restrictive temperatures. No accumulation of precursor envelope polyprotein in the CNS was evident in the infected mice. This suggests that *ts* Mo BA-1 may employ a different mechanism to cause neuronal degeneration.

Two of the paralytogenic *ts* mutants, *ts*1 and *ts*7, together with MoMuLV-TB were molecularly cloned. All of the molecularly cloned mutants, five clones from *ts*1 and one from *ts*7, were found to have retained the characteristics of their nonmolecularly cloned parents (Yuen et al. 1985a). These characteristics are: temperature sensitivity, inefficiency in the intracellular processing of $Pr80^{env}$ at the restrictive temperature, enhanced neurotropism, and ability to induce paralysis in susceptible mice. These studies further establish that these paralytogenic *ts* mutants are the bona fide etiologic agent for the paralytic disease in the susceptible host.

The three paralytogenic mutants of MoMuLV-TB differ in their efficiency to induce paralysis in susceptible host. *Ts*1, the prototype of these mutants, is the most effective inducer of paralysis in susceptible mice. The majority of *ts*1-infected CFW/D mice, when inoculated intraperitoneally within 48 h after birth, became paralyzed by 40 days p.i. whereas the majority of *ts*7-infected mice became paralyzed by 40–60 days p.i. (Yuen et al. 1985a). The latency period for mice infected with *ts*11 is similar to that of *ts*7 (P.K.Y. Wong, unpublished data). Both *ts*7 and *ts*11 in addition to defective processing of the $Pr80^{env}$ precursor protein are also heat labile (Wong and Gallick 1978 and unpublished data). This latter defect distinguishes *ts*7 and *ts*11 from *ts*1. The additional defect (i.e., heat lability) of *ts*7 and *ts*11 may account for the more protracted latency period of paralysis induced by these mutants. Among these three paralytogenic *ts* mutants of MoMuLV-TB, *ts*1 is the most extensively studied; therefore, the major portion of this review is based on studies of *ts*1.

3 Host Strain Susceptibility

Mouse strains vary considerably in their susceptibility to the neurologic disease induced by *ts* mutants of MoMuLV-TB (Table 2). CFW/D mice were chosen for initial studies of the pathogenesis of the CNS disease (McCarter et al. 1977;

Table 2. Comparison of the incidence and development of neurologic disorders in different strains of mice induced by *ts*1 MoMuLV-TB and *ts* Mo BA-1 MuLV

Strains	Percentage with neurologic disorders		Onset range (weeks)	
	*ts*1[a]	*ts* Mo BA-1[b]	*ts*1[a]	*ts* Mo BA-1[b]
CFW/D	100	NT	4–7	
FVB	100	NT	4–5	
BALB/c	98	80	4–10	10–20
C3H	98	NT	6–10	
SWR/J	100	NT	7–10	
CBA	82	80	8–12	10–20
NIH/Swiss	90	NT	8–12	
NFS	NT	98		
SJL	NT	80		10–20
C57BL	0	0		
B10.D2	0	NT		
SENCAR	0	NT		

[a] P.K.Y. Wong (unpublished data)
[b] Bilello et al. 1986
NT, not tested

Wong et al. 1983, 1985; Yuen et al. 1985 a, b; Zachary et al. 1986) because they are highly susceptible to the neurologic disease induced by the *ts* paralytogenic mutants. This could be due to the fact that MoMuLV-TB and its mutants have been adapted to CFW/D mice. Since their isolation they have been propagated and passaged in TB cells, a thymus/bone marrow cell line which was derived from CFW/D mice (Ball et al. 1964). As mentioned above, among the three paralytogenic mutants of MoMuLV-TB, *ts*1 appears to be most potent. Four weeks after inoculation with *ts*1, over 80% of the CFW/D mice developed clinical symptoms of paralysis and histopathologic evidence of CNS disease (Zachary et al. 1986). In contrast, none of the CFW/D mice inoculated neonatally with MoMuLV-TB or MoMuLV developed clinical symptoms of paralysis. Results in CFW/D mice were later duplicated in other susceptible inbred strains (Table 2) (McCarter et al. 1977; Wong et al. 1983; E. Floyd and P.K.Y. Wong, unpublished data). Among these susceptible strains, the age of onset of paralysis is not uniform and generally exhibits a longer latency period than in CFW/D mice. Of these strains of mice, the FVB is the most susceptible, which developed hind limb paralysis about 4–5 weeks p.i. This is followed by BALB/c, C3H, NSF, and SWR/J which developed paralysis about 5 to 10 weeks p.i. CBA and NIH/Swiss are less susceptible and they developed paralysis about 8–12 weeks p.i. In contrast, C57BL and B10.D2 mice, when inoculated with *ts*1 do not develop paralysis (Table 2). Rats are not susceptible to *ts*1-induced paralysis (McCarter et al. 1977). However, virus obtained from C57BL mice or rats infected with *ts*1 retains the ability to induce paralysis in susceptible mice. Since both the BALB/c strain, which is susceptible, and the B10.D2 strain, which is resistant to *ts*1-induced paralysis, carry the same H-2d gene complex, this indicates that the major mouse histocompatibility complex (H-2) locus is not a critical determinant of susceptibility but other host factors may play a critical role in the susceptibility to the paralysis induced by *ts*1.

In the *ts* Mo BA-1 system, BALB/c, C3H, CBA, NFS, and SJL mice are susceptible whereas C57BL mice are resistant to the neurologic disease. These data are, in general, consistent with the strain suceptibility reported for *ts*1 MoMuLV-TB except in *ts* Mo BA-1 the latency period is lengthened (about 10–20 weeks p.i.) and the disease frequency is reduced (Table 2).

Interestingly, the mouse strain susceptibility to *ts*1 and *ts* MoBA-1 is generally in agreement with that reported for WM-E virus-induced neurogenic paralysis (Oldstone et al. 1977; Brooks et al. 1980) except for BALB/c mice, which are resistant to the WM-E virus-induced paralysis. For example, C3H, SWR/J, and NIH/Swiss mice are susceptible whereas C57BL and B10.D2 mice are resistant to both virus systems. The main differences between the *ts*1 and WM-E virus host systems are that BALB/c mice are susceptible to the former and resistant to the latter system. This difference is mainly due to the fact that MoMuLV is NB-tropic whereas WM-E virus is N-tropic. Because MoMuLV is NB-tropic, no genetic restriction for virus replication is conferred by the mouse *Fv-1* locus (Pincus et al. 1971) to the *ts* mutants of MoMuLV, resulting in unrestricted replication of *ts*1 in BALB/c mice, in particular in the CNS, and as a result of viral replication neurologic disease is generated. This is also consistent with the observations in both the WM-E virus and *ts*1 MoMuLV-TB systems that susceptible strains of mice show consistently higher virus titers in their sera and also in their tissues, particularly in the CNS, than resistant strains of mice such as C57BL (Oldstone et al. 1977; Wong et al. 1985; P.K.Y Wong and E. Floyd, unpublished data).

Attempts have been made in the author's laboratory to determine whether immune system factors are responsible for resistance of C57BL mice to hind limb paralysis induced by *ts*1. Specific immune effector functions were deleted by using anti-thy1.2 and cyclosporin A for T cells, anti-μ and cyclophosphamide for B cells, silica for macrophages, and anti-gamma interferon for the interferon system. These treatments were carried out shortly before the injection of neonate C57BL mice with *ts*1 and for 10 consecutive days after inoculation. These treatments did not render C57BL mice susceptible to *ts*1-induced neurologic disorders, suggesting that immune factors may not be responsible for the manifested resistance of C57BL mice to *ts*1. Since C57BL mice inoculated with *ts*1 neonatally show consistently lower virus titers throughout the course of infection in their sera and in their tissues than susceptible strains of mice such as CFW/D and BALB/c mice, perhaps resistance of C57BL to *ts*1 is related to specific virus-host interactions which affect the level of virus expression in specific cells, in particular cells associated with CNS tissue (E. Floyd; G. Prasad and P.K.Y. Wong, unpublished observation).

Studies with a number of other neurotropic viruses using inbred strains of mice with well-defined genetic backgrounds have shown that host genes, acting through a variety of mechanisms, can influence the resistance and susceptibility of the host to neurotropic viruses (for review see Tyler 1988). To further determine the host genetic control of resistance and susceptibility in the *ts*1 system, studies with recombinant inbred strains between genetically well defined resistants and susceptible strains are necessary.

4 Clinical Features

*Ts*1-inoculated susceptible mice show a predictable clinical and histologic profile. When CFW/D mice were inoculated within 24 h of birth, mild generalized body tremors are detected as early as 21 days p.i., which progresses to a notable bilateral hind limb paresis by 28–35 days p.i. (McCarter et al. 1977; Wong et al. 1983; Zachary et al. 1986). As pointed out by Stansly (1965), the retraction of the hind limbs is a useful diagnostic sign of early paralysis; this is particularly striking when the mice are suspended by their tails. As the disease progresses, hind limb positioning and movement changes from one of abduction about the pelvic joint to one of adduction and voluntary movement in all joints in the hind limb become reduced (Zachary et al. 1986). By 28–35 days p.i., paralysis develops and hind limb movement is limited to the tibial-tarsal joint. Kyphosis and frequent intention tremors are observed. From this point on, *ts*1-infected mice deteriorate rapidly. Tremors increase in severity and can be elicited with any disturbance. Mice have no interest in eating or drinking; however, all have food in their stomachs at necropsy. Death usually occurs by 32–40 days p.i. and is generally preceded by spontaneous tremors and obvious signs of dyspnea.

Beside neurologic signs, *ts*1-infected mice also exhibit some other symptoms. *Ts*1-infected mice are usually scruffy and runted. While mice normally show rapid growth and weight gain between 20 and 30 days of age, mice infected with *ts*1 show no weight gain after 20 days of age (C. Knupp and P.K.Y. Wong, unpublished data). Some infected mice also show symptoms of diarrhea, pneumonia, and eye infections which are often associated with secondary infection due to immunologic dysfunction. Necropsy revealed severe thymic atrophy in most of the paralyzed mice (McCarter et al. 1977). Reduction of the size of the spleen was also observed (C. Knupp and P.K.Y. Wong, unpublished data). In addition, hind limb musculature appeared pale and reduced in size when compared with that of normal control mice. No gross lesions are observed in other organs of *ts*1-infected mice at necropsy. Essentially similar clinical features and symptoms were observed in BALB/c mice infected with *ts*1 within 48 h after birth (Wong et al. 1989). However, the time of onset and progression of the disease from onset to paralysis and death is more variable in BALB/c than in CFW/D mice and generally is more protracted (see section on host strain susceptibility). Since in *ts*1-infected BALB/c mice paralysis is generally delayed and the mice survive longer, symptoms such as body wasting, severe atrophy of thymus and spleen, and secondary infection, in addition to those associated with neurologic disorders, are more apparent than in *ts*1-infected CFW/D. Because some of these symptoms are characteristics of immunosuppression in the infected host, a detailed study of the effect of *ts*1 on the immune system of the BALB/c mice has been recently performed in the author's laboratory (Wong et al. 1989). Some of the data of this investigation will be presented in a later section of this review.

Bilello and coworkers observed that neonatal mice inoculated with *ts* Mo BA-1 virus became mildly tremulous at about 8–10 weeks p.i. and progressed to a moderate degree of tremor and inability to extend the hind limbs. Rare instances of hind limb paralysis were observed in the later stages of the disease (Bilello et al. 1986). The clinical presentation of *ts* Mo BA-1 MuLV-induced neurologic disease appears to be milder than that of *ts*1 MoMuLV-TB, WM-E virus as well as rat-passaged F-MuLV and represents another class of ecotropic MuLV that is capable of inducing neurodegenerative disease.

5 Age-Related Resistance to Induction of Paralysis

Experimental infection of CFW/D mice with *ts*1 at varying times during the first 10 days after birth has a pronounced effect on the time of onset and rapidity of progression of hind limb paralysis. Mice injected within the first 48 h after birth showed the first signs of symptoms of paralysis at approximately 4–5 weeks of age and lived an average of 5–6 days thereafter. Mice injected on days 3,4, and 5 postpartum developed paralysis at times ranging from 7 to 10 weeks postinfection and survived considerably longer in a state of slowly progressing paralysis. Mice injected on the 10th day after birth appeared normal until 8–10 months of age. These latter animals maintained a more chronic and less degenerative paralytic condition for 1–3 months before succumbing to the disease (B. Springer and P.K.Y. Wong, unpublished data). The viremic titers of CFW/D mice injected with *ts*1 on day 1, day 5, or day 10 postpartum is presented in Fig. 2. The ability of the virus to replicate and give rise to viremia is related to the age of the host at infection. It is apparent that the virus titer in the blood increases much more slowly in animals inoculated later than day 1 postpartum. The later the

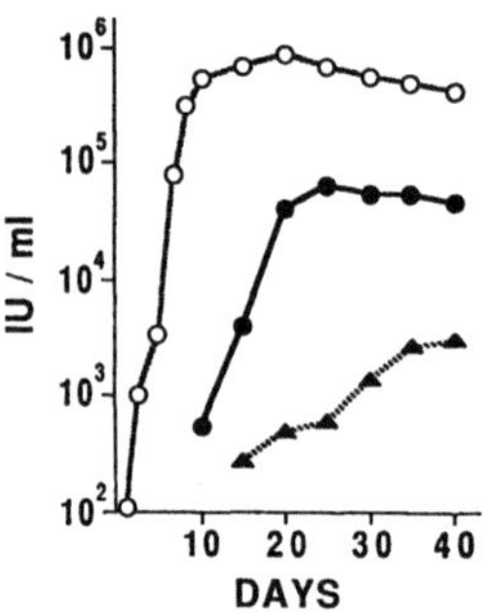

Fig. 2. Rate of development of viremia in CFW/D mice injected with the *ts*1 mutant virus on days 1, ○; 5, ●; and 10, ▲ postpartum

virus is introduced, the longer it takes the animals to develop viremia. This results in a delayed disease onset and more insidiously progressive paralysis.

Similar to CFW/D mice, resistance to *ts*1-induced paralysis in BALB/c mice develops as the mice mature. However, in BALB/c mice, unlike CFW/D mice, paralysis can no longer be induced if mice are injected later than 5 days of age (G. PRASAD and P.K.Y. WONG, unpublished observation). These observations relating to age-related resistance to induction of paralysis are consistent with those observed in the WM-E virus system (BROOKS et al. 1981; HOFFMAN et al. 1981).

Age-related resistance factors which may be important in determining whether paralysis develops following viral inoculation includes development of the immune system of the mouse, loss of susceptible target cells in the CNS with age, and differentiation or maturation of neurons and glial cells (BROOKS et al. 1981) which may render them less susceptible to viral infection.

BALB/c mice inoculated on day 10 with *ts*1 had evidence of infection including viremia and presence of infectious virus in their spleen and CNS. However, titers in the sera, spleen, and CNS of these mice were significantly lower than those of the paralyzed mice inoculated after birth. Their sera contained neutralizing antibodies against MoMuLV, whereas sera from BALB/c mice inoculated within 48 h after birth did not. This study indicates that *ts*1 virus, which replicates well in neonate BALB/c mice, becomes persistent either because of neonatal tolerance to the virus or because the virus burden overwhelms the immature immune response of the newborn mice. Ten-day-old BALB/c mice, however, are capable of making a significant immune response against *ts*1 virus (G. PRASAD and P.K.Y. WONG, unpublished data).

The presence of anti-MoMuLV antibodies in the sera of BALB/c mice inoculated at 10 days of age with *ts*1 which did not develop paralysis prompted this author and his coworkers to investigate the role of the immune system in age dependent resistance to *ts*1-induced paralysis in BALB/c mice. Specific immune functions were suppressed as detected with anti-T-cell serum for T cells and cyclophosphamide for B cells in BALB/c mice inoculated with *ts*1 at 10 days of age. None of these treated mice developed neurologic disorders. Transfer of splenic and thymic lymphocytes from young adult BALB/c mice to neonatal BALB/c pups 4 h prior to *ts*1 inoculation did not prevent hind limb paralysis. The results of these studies suggest that the immune system may not be responsible for age-dependent resistance to hind limb paralysis. Furthermore, BALB/c nude mice injected even within 24–48 h of birth did not develop neurologic disorders. This further supports the observation that at least T-cell functions are not responsible for the age-related resistance to paralysis (PRASAD et al. 1989). These observations, however, are in contrast to those observed by HOFFMAN and coworkers (1984) in their studies on the age-related resistance of mice to paralysis induced by WM-E virus. They concluded that resistance was due to antiviral immune response.

If the development of the immune system, including mature T cell function, is not responsible for the age-dependent resistance to paralysis induced by *ts*1, the maturation of neurons and other nerve cells in the CNS may play a critical role. It is known that retrovirus replication generally requires one cycle of cell division and DNA synthesis for integration of proviral DNA. Since neuronal division in mouse CNS ceases early in the postnatal period (UZMAN and RUMLEY 1958) to establish neuronal infection, *ts*1 virus must gain access to the CNS tissue during the first few days of life.

6 Virus Replication in the Infected Host

*Ts*1 virus can be transmitted to the host by perinatal transfer from mother via milk to neonate (P.K.Y. WONG and S. ROSS, unpublished data). In infection with mechanically transmitted virus, the route of injection, i.e., intracerebrally (i.c.) or intraperitoneally (i.p.), is not crucial except that neonatal mice inoculated i.c. take a slightly longer time to develop the first signs of paralysis (R. MACLEOD and P.K.Y. WONG, unpublished data). Following i.p. inoculation of neonate mice with *ts*1, virus replicates primarily in the thymus and spleen and establishes a systemic infection with viremia. Within 5 days p.i., considerable amounts of infectious virus can be detected in the thymus. The concentration of infectious virus recovered from the thymus increases rapidly and reaches a peak by 10–15 days p.i. Virus replication in the thymus is followed by replication in the spleen. The concentration of infectious virus recovered from the spleen reaches a plateau by 15–20 days p.i. Virus replication in the lymphoid organs is followed by viremia and the virus then spreads to other organs including the CNS (WONG et al. 1985; PRASAD et al. 1989).

To gain access to the CNS, the virus must penetrate the blood-brain barrier. The mechanism(s) by which *ts*1 viral penetration of the blood-brain barrier is not clear, but studies on other virus systems provide some clues. It was suggested by KOEING et al. (1986) that HIV infects the CNS via macrophages which act as carriers and penetrate the blood-brain barrier by diapedesis through capillaries in the CNS. Another mechanism implicated for other neurotropic RNA viruses is that some viruses infect axon terminals in skeletal muscles, with subsequent transportation up the spinal cord, and finally infect the lower motor neurons in the spinal cord and brain stem (KRISTENSSON and NORBY 1986). SWARZ et al. (1981) suggested that the WM-E virus enters the CNS by replicating in capillary endothelial cells and budding from the endothelial cells into the CNS. This hypothesis was substantiated by BILELLO et al. (1986), who reported that CNS capillary endothelial cells and pericytes appear to be the earliest replication site for *ts* Mo BA-1 virus infection.

*Ts*1 virus must gain access to the CNS during the first few days of life (see section on age-related resistance to induction of paralysis). Once inside the CNS, the virus infiltrates specific nerve cells and replicates in those cells. The replication of *ts*1 virus in the spinal cord and brain appears to be slower than that found in the thymus and spleen. However, virus titers in the spinal cord and brain increase gradually and exceed that in the plasma by 25–35 days p.i., when paralysis becomes apparent. At this time, the virus titer in the CNS of *ts*1-infected mice is about 100-fold higher than that in the CNS of MoMuLV-TB-infected mice. The virus titer also correlates with the severity of the lesion found in the CNS (WONG et al. 1985; ZACHARY et al. 1986). The infectivity detected in the *ts*1-infected CNS is not due to the presence of endogenous viruses, since the infectious virus recovered possesses the phenotype of the *ts*1, i.e., temperature sensitivity and nonprocessing of $Pr80^{env}$. These studies indicate that *ts*1 replicates in the CNS, especially in the spinal cord, much more efficiently than does the parental MoMuLV-TB virus. This enhanced neurotropism of *ts*1 virus is further indicated by the observation that *ts*1 grows to a much high titer in primary cultures of neurons than MoMuLV-TB (WONG et al. 1985).

The seeming paradox that *ts*1 virus replicates better than its wild-type counterpart in the CNS tissues in the infected mice, despite its temperature sensitivity and reduced ability to process $Pr80^{env}$, was explained by WONG et al. (1985) as follows: the body temperature of newborn and young mice (1–20 days old) is approximately 34°C (the permissive temperature for *ts*1 replication) and that of adult mice (more than 3 weeks old) is approximately 38.4°C (CRISPENS 1978). This latter temperature is nonpermissive for *ts*1 replication. It therefore appears that the replication of *ts*1 in young mice is not restricted and *ts*1 virus could spread to the CNS and establish infection in the nerve cells. After 20 days of age, as the mice achieve adulthood, their body temperature may become more restrictive to the ability of *ts*1 to process $Pr80^{env}$ to gp 70 and Prp15E, which may affect the amount of infectious virus produced (i.e., affect its ability to spread from cell to cell). However, in mice older than 20 days, the ability of *ts*1 virus to spread from cell to cell may not be as critical for the induction of paralysis because most of the cells are probably infected before 20 days of age.

In contrast, no enhanced neurotropism has been observed in *ts* Mo BA-1 virus. At 12 weeks p.i., the amount of detectable infectious virus in the nervous system tissue to *ts* Mo BA-1-infected mice demonstrating neurologic disorders is similar to that detected in wt Mo-MuLV-infected mice that are asymptomatic. Furthermore, in *ts* Mo BA-1-infected mice, viral proteins are localized primarily in spleen megakaryocytes and in vascular endothial cells of the cerebellum and spinal cord and hardly any are detected in the neurons (BILELLO et al. 1986). BILELLO et al. concluded that the mechanism responsible for neurodegeneration appears to require more than the ability of the virus to replicate in the CNS. However, the additional factor(s) which is responsible for *ts* Mo BA-1 virus to induce neurodegenerative disease has not been resolved. PITTS et al. (1987) observed ultrastructural changes in capillary endothelial cells of the CNS of *ts* Mo BA-1-infected mice and postulated that these ultrastructural changes in endothelial cells may be associated with the progressive development of the spongiform lesions. However, whether similar ultrastructural alterations of capillary endothelial cells also occur in MoMuLV-TB-infected mice has not been studied.

7 Histopathologic Studies

A progressive symmetric noninflammatory spongiform encephalomyelopathy was observed in the brain and spinal cord from CFW/D mice inoculated with *ts*1 (McCarter et al. 1977; Wong et al. 1985; Zachary et al. 1986). The histologic lesions observed in *ts*1-infected mice are in most respects similar to those reported in studies with the *ts* Mo BA-1 MuLV (Bilello et al. 1986), the wild mouse neurotropic retrovirus models (Andrews and Gardner 1974; Brooks et al. 1980; Gardner et al. 1973; Oldstone et al. 1977, 1980, 1983), the rat-passaged paralytogenic F-MuLV (Kai and Furata 1984), the twitcher mouse model (Kobayashi et al. 1988), and the slow virus diseases such as scrapie in sheep, as well as scrapie and Creutzfelt-Jakob disease of man (Lampert et al. 1972). The noninflammatory histologic lesions induced by *ts*1 MoMuLV-TB are also to some extent similar to the HIV-related vacuolar myelopathy found in some AIDS patients (Anders et al. 1986).

The severity and progression of neurodegenerative signs in *ts*1-infected mice correlate with the location, development, and progression of lesions in the CNS. Lesions in the CNS are symmetric, increase in severity with time, and consistently arise at specific times in specific nuclei and areas of the brain and spinal cord (Zachary et al. 1986). The onset and severity of the spongiform changes also correlate with maximum titers of *ts*1 virus in the CNS (Wong et al. 1985; Zachary et al. 1986). Lesions appeared in the cerebral cortex, brain stem, and cervical as well as lumbar spinal cord simultaneously. In general, lesions in the lumbar region were more severe than in the cervical spinal cord, and lesions in brain stem-cerebellum were more pronounced than in the spinal cord. Cerebellar nuclear lesions were more prominent than brain stem nuclear lesions at early time points, but brain stem nuclear involvement became greater with time (Zachary et al. 1986). The histologic observations in *ts*1-infected mice are somewhat different from those in *ts* Mo BA-1-infected mice. In *ts* Mo BA-1-infected mice, less severe changes were noted in the cerebrum and lower spinal cord and degenerating neurons were mostly observed in the cerebellum, brain stem, and anterior horn of the upper spinal cord (Bilello et al. 1986).

The reason for the greater severity of lesions in the lower spinal cord in *ts*1-infected mice is unclear. This may be related to the mode of virus spread to the CNS, or to the susceptibility of specific nerve cell populations to infection by *ts*1, since a relatively higher virus titer was observed in the lumbar spinal cord than in the cervical spinal cord (P.K.Y. Wong and K. Knupp, unpublished data). The greater severity of lesions in the lower spinal cord may account for the dominance of hind limb involvement in *ts*1-infected mice. A greater concentration of spongiform changes was also noticed in the lower spinal cord in WM-E retrovirus-infected mice which predominately manifest paralysis of the hind limbs. In contrast, in *ts* Mo BA-1-infected mice, greater severity is found in the cervical than the lumbar spinal cord. This may explain why hind limb paralysis is rarely observed in *ts* Mo BA-1-infected mice (Bilello et al. 1986).

Histologic findings confirmed the neurologic signs of flaccid paralysis caused by lower motor neuron disease in *ts*1-infected mice. Lateral and vertical funiculi had generalized symmetric spongiform changes and, in some instances, lesions were most pronounced at the gray/white matter junction. In addition to the gray matter and the gray/white matter junction, spongiform changes were also observed in the white matter in the *ts*1-impaired CNS (ZACHARY et al. 1986). This latter observation differs somewhat from those observed in the CNS of neurologically impaired mice infected with *ts* Mo BA-1 MuLV (BILELLO et al. 1986) or WM-E virus (OLDSTONE et al. 1977, 1980; SWARZ et al. 1981), whereby spongiform changes were mainly confined to the gray matter and the gray matter/white matter junction. However, SMITH et al. (manuscript submitted) have noted white matter involvement in molecularly cloned WM-E virus (Cas Br-M)-infected mice.

The distribution of spongiform changes in the CNS of *ts*1-infected mice suggests that, in addition to involvement of lower motor neurons, major involvement of upper motor neurons of the extrapyramidal system also occurs. It is possible that neurologic signs of upper motor neuron disease were not observed in *ts*1-infected mice because they were masked by a combination of lower motor neuron disease and cerebellar nuclear lesions (ZACHARY et al. 1986).

In the brain and brain stem of neurologically impaired CFW/D mice infected with *ts*1, lesions are observed in nuclei of the extrapyramidal system, nuclei of the brain stem with integrative or associative motor functions, nuclei with major sensory functions, and nuclei of cranial nerves. It is not known why, in general, cerebellar nuclear lesions were more pronounced in the early stages of the disease than were brain stem nuclear lesions, whereas the latter were greater in severity than all other areas of the CNS in the later stages of the disease. Nor is it known why brain stem-cerebellar lesions are more prominent than spinal cord lesions. As mentioned above, these findings may be related to viral tropism or to the susceptibility of specific neuron populations to virus-induced damage. Preliminary ultrastructural studies by ZACHARY and associates (1986) indicated that funicular spongiosis might involve granular dissolution of the axoplasm and myelin loss with separation of myelin lamellae. Whether myelin loss was secondary to axonal degeneration as a result of viral-induced damage to oligodendroglial cells, or due to a combination of both mechanisms, has not been resolved. Wallerian degeneration, separation of myelin lamellas, redundant myelin sheaths, neuronal and oligodendrocyte degeneration, and granular disintegration of axoplasm have been reported in ultrastructural studies with the wild mouse paralytogenic retroviruses (ANDREWS and GARDNER 1974; OLDSTONE et al. 1977; SWARZ et al. 1981). These authors cosnidered that both neuronal and oligodendrocyte viral-induced degeneration were responsible for these changes.

Lesions in the nuclei of the reticular formation and the hypothalamus were extensive and severe. It is plausible that lesions in these nuclei may have ultimately led to the death of *ts*1-infected mice due to dysfunction of respiratory, conscious, or vasomotor centers. Lesions in the thalamic nuclei and the rostral and caudal colliculi suggest involvement of sensory systems in the absence of lesions in posterior gray horn and dorsal funiculi of the spinal cord (ZACHARY et al. 1986). These findings differ in some aspects from other murine neurotropic retrovirus systems (GARDNER 1985; BROOKS et al. 1980; OLDSTONE et al. 1977).

It is not known why lesions would favor localization in some CNS regions rather than others. This specificity may be regulated in part by surface receptors for virions that are common between neuronal and glial cell populations (GARDNER 1985; BROOKS et al. 1981). It is plausible, however, that specificity is determined by endothelial cell surface receptors and the distribution of neural and glial cell lesions is a reflection of this specificity (PITTS ET AL. 1987). In addition, endothelial cells and glial cells of the brain may possibly serve as important sites for viral integration and replication in a dividing-cell population, in contrast to the postmitotic status of neurons. As mentioned earlier, the mechanism of how *ts*1 virus spreads to the CNS after intraperitoneal or intracerebral inoculation is not clear. Other workers using *ts* Mo BA-1 MuLV and wild mouse neurotropic retroviruses (PITTS et al. 1987; SWARZ et al. 1981) have demonstrated that during viremia the virus infects capillary endothelial cells of the CNS and spreads contiguously into the neuropil by budding from the abluminal surface of the endothelium through the basement membrane. Infection of specific glial and neuronal cells apparently follows shortly thereafter. Type-C virus budding from the abluminal surface kf the endothelium and aligning along or within the basal laminae was also observed in the capillary endothelial cells of the CNS of *ts*1-infected mice (C. KNUPP and P.K.Y. WONG, unpublished data). The number of virus particles observed in this area is much less than those reported by PITTS and coworkers (1987) in the *ts* Mo BA-1-infected mice. Therefore, whether *ts*1 also uses a similar mechanism to spread and impair the CNS is not yet resolved in detail.

8 Ultrastructural Studies

Ultrastructural studies revealed typical type-C viruses budding from the plasma membrane of virtually all cell types in the bone marrow of *ts*1-infected CFW/D mice as early as 5 days p.i. Bone marrow and splenic megakaryocytes and platelets showed massive viral accumulation in the cytoplasmic cisternae and occasionally in rhabdoform assembly (R. MACLEOD and P.K.Y. WONG, unpublished observation). Budding virus was also observed on thymic reticular epithelial cells, thymocytes, and in lymphocytes of the splenic white pulp. Aggregates of virus were observed on the cytoplasmic processes of the dendrites of reticular cells in the lymph node cortex. Virus was observed budding from endothelial cells and as free particles in the lumina of the majority of the vasculature. Examination of the kidney revealed aggregates of virus lodged in the basement membranes of the glomeruli. No local thickening of the basement membrane nor fusion of the foot processes which are often associated with immune-complex disease was observed. Virus was also observed in the hind limb musculature.

The CNS of *ts*1-infected CFW/D mice during the later course of infection revealed viral particles in extracellular space, associated with the Golgi complex and plasmalemma of glial cells and budding from neuronal plasmalemma (ZACHARY et al. 1986). Degenerative changes observed in neurons and glial cells suggested that they are most likely caused by viral infection in these cells. This conclusion is in general agreement with findings in the wild mouse virus models (ANDREWS and GARDNER 1974; GARDNER 1985). Caution must be used, however, in interpretation of the *ts*1 system because, in contrast to the wild mouse neurotropic retrovirus systems, *ts*1 viral particles were not found in the endoplasmic reticulum or Golgi complex or budding aberrantly into cytoplasmic vacuoles of neurons (MCCARTER et al. 1977; ZACHARY et al. 1986; R. MACLEOD and P.K.Y. WONG, unpublished data). The intimate association between glial cells, especially astrocytes, and neurons and the role of astrocytes in regulating the neuronal environment suggested that neuronal degeneration may be secondary to astrocyte dysfunction. It is likely that viral infection of neurons and glial cells affected cellular metabolism, most significantly protein synthesis necessary for normal membrane function and ion transport since in *ts*1-infected mice neurons had the most pronounced dilatation of endoplasmic reticulum and Golgi complex of all cells in the CNS. This could be a primary change caused by neuronal infection with *ts*1 or a secondary change due to dysfunction of supporting glial cells or virus. The latter hypothesis, however, is not likely because experimental and spontaneous diseases with severe dysfunction of astrocytes are, in general, not associated with dilatation of neuronal endoplasmic reticulum or Golgi complex (ADACHI et al. 1973; ZACHARY and O'BRIEN 1985). This dilatation could be explained by the following hypothesis: *Ts*1 virus is defective in processing the precursor envelope polyprotein $Pr80^{env}$ to gp70 and Prp15E, and it is plausible that accumulation of the precursor envelope protein causes dilatation of the endoplasmic reticulum and Golgi complex. $Pr80^{env}$ may be inserted into the membrane of the endoplasmic reticulum and Golgi complex or improperly processed after insertion, resulting in excessive formation of membrane. Lack of processing may also explain why such low numbers of virus particles are seen in neurons and glial cells. It is also plausible that $Pr80^{env}$ binds to the cytoskeleton of the neurons. This interaction may influence axoplasmic transport. Further studies are needed to define the role of $Pr80^{env}$ in the dilatation of the neuronal and glial endoplasmic reticulum and Golgi complex and its influence on cell degeneration. Neuronal degeneration in the *ts*1 system is a slow process that results in a progressive cumulative dysfunction of neurons. This injury is apparently irreversible and ultimately results in neuronal death and loss.

Work with the wild mouse neurotropic retrovirus models has produced hypotheses that suggest neuronal injury might be due to productive intraneuronal retrovirus replication and accumulation in the cytoplasmic vacuoles (OLDSTONE et al. 1980; SWARZ et al. 1980), to viral-derived proteins toxic to neurons (GARDNER 1978), to alterations in cyclic nucleotide metabolism (BROOKS et al. 1983), or to interference with normal neuron metabolism (BROOKS et al. 1980). An alternative to the hypothesis that the neuronal degeneration results from productive intraneuronal virus replication was proposed by PITTS and coworkers (1987). Based on their observations that widespread ultrastructural changes associated with viral replication in the vascular endothelial cells in the CNS of symptomatic mice infected with three different neuropathogenic MuLVs, including *ts* Mo BA-1, PITTS et al. (1987) suggested that these ultrastructural changes may alter normal physiologic functions and, thus, could play a significant role in the pathogenesis of the spongiform changes observed in the CNS.

Histologic neuronal and glial lesions were also observed in CFW/D mice infected with wt MoMuLV-TB virus (ZACHARY et al. 1986). However, these lesions were not as severe as those in *ts*1-infected mice and regressed as the study progressed without apparent loss of neurons. MoMuLV-TB does synthesize $Pr80^{env}$; however, it is efficiently cleaved to gp70 and p15E. It is possible that early in infection with MoMuLV-TB virus enough $Pr80^{env}$ is synthesized to cause histologic changes without functional abnormalities that result in neurologic signs. MoMuLV-TB-infected mice appear able to resolve the injury by converting $Pr80^{env}$ to gp70 and Prp15E and possibly minimize the hypothesized dysfunction of cellular organelles or cytoskeleton associated with excessive early synthesis of $Pr80^{env}$. The difference in the progression of lesions between MoMuLV-TB and *ts*1 may also be due to the fact that MoMuLV-TB mainly infects glial cells but rarely infects neurons (WONG et al. 1985) and that damaged glial cells could be repaired replenished whereas damage to neurons is more permanent.

9 Involvement of the Immune System in Neurologic Disorders

Despite the fact that susceptible mice injected within 48 h after birth showed paralysis as early as 30–35 days, the possibility existed that the mutant virus, having tropism for the cells and tissues of the immune system, could modify the behavior of the cells of the immune system, perhaps through precocious maturation. Moreover, the observation that the level of immunoglobulin in the plasma of paralyzed animals was elevated and that congenital athymic (nude) mice are resistant to paralysis induced by *ts*1 (PRASAD et al. 1989) lends credibility to this hypothesis. However, histologic examination of the neural tissues from severely paralyzed mice showed no signs of mononuclear cell infiltration into the areas of the lesions, indicating that immune system cells may not be directly associated with neuronal damage through an attack on CNS cells infected with the virus.

An approach to determine whether in *ts*1-paralyzed mice a humoral immune response to the virus has developed resulting in autoimmunity was performed by examining immune aggregates in frozen sections (F. MACLEODE and P.K.Y. WONG, unpublished data). This study showed that while the kidneys from all of the paralyzed mice revealed an abundance of mature virus particles in and around the basement membrane of glomeruli and strongly positive fluorescence in this area when stained with anti-Mo-MuLV antibodies, they were consistently negative for the presence of glomerular immune-aggregate deposition when stained with fluorescein-conjugated antimouse immunoglobulin antibodies. Furthermore, chromium release experiments using both healthy and infected splenocytes as targets, and plasma from paralyzed mice with guinea pig complement as effectors, were negative for complement-mediated antibody cytotoxicity. These observations together with the fact that no anti MoMuLV antibodies were detected in the sera of paralyzed mice infected with *ts*1 neonatally (G. PRASAD and P.K.Y. WONG, unpublished data) suggest that neither antiviral nor autoreactive antibodies were involved in the *ts*1-associated paralysis. These observations are in agreement with the conclusions of GARDNER et al. (1973) and OLDSTONE et al. (1977) that damage to the nervous system in the WM-E virus system does not involve the immune system.

10 Role of Thymus in the Pathogenesis of Hind Limb Paralysis

The thymus and spleen, in particular the T-cell areas, are the primary target of and major organs for virus production in *ts*1-infected BALB/c mice (WONG et al. 1989; PRASAD et al. 1989). For this reason the thymus may play an important role in the induction of paralysis by *ts*1. To investigate the role of T cells or the thymus in the paralysis induced by *ts*1, several studies have been initiated in the author's laboratory recently. One of these studies is to use homozygous (nu/nu) athymic nude mice to evaluate the role of T-cells in *ts*1-induced paralysis. Results obtained from these studies indicate that athymic BALB/c nu/nu mice are resistant to the neurologic disorders induced by *ts*1 (PRASAD et al. 1989). Furthermore, preliminary studies with adoptive transfer of T cells to nude mice indicate that T-cell-reconstituted nude mice are susceptible to *ts*1-induced paralysis. These observations strongly suggest that the thymus plays an important role in *ts*1-induced hind limb paralysis. The observation that nude mice are resistant to *ts*1-induced paralysis is different from that of WM-E virus observed by OLDSTONE and coworkers (1977), who reported that nude mice infected with WM-E virus developed paralysis. The differences observed between *ts*1 MoMuLV-TB and WM-E virus systems could be explained by the fact that *ts*1 MoMuLV-TB is T-cell tropic whereas WM-E virus is B-cell tropic. In the latter system, virus replication was restricted in susceptible mice made asplenic with cyclophosphamide (BROOKS et al. 1980) and paralysis can be prevented by splenectomy of newborn mice before WM-E virus inoculation (GARDNER 1979).

11 Immunologic Abnormalities Induced by *ts*1 in Infected Mice

While earlier in vivo studies have emphasized the CNS disease induced by *ts*1, a systematic examination of *ts*1 pathology in other organs, in particular the thymus and spleen, has recently begun in the author's laboratory (WONG et al. 1989). Histologic examination of spleen and thymus of *ts*1-infected BALB/c mice 30–50 days p.i. revealed massive depletion of lymphocytes in the thymus and mild depletion of lymphocytes from the T-cell areas in the spleen resulting in extensive thymic and splenic atrophy although the latter is not as severe as the former. It appears that *ts*1 is cytopathic for T-lymphocytes, leading to depletion of lymphoid areas in both organs, although the effect in the thymus is more severe. Depletion of lymphocytes in the infected mice is also reflected in the peripheral blood count. By 40 days p.i. there is a drastic reduction in the total lymphocyte count, indicating lymphopenia in the infected mice. In vitro studies indicated that the ConA-induced lymphoproliferative response of T cells obtained from *ts*1-infected mice was drastically suppressed. In addition to T cells, *ts*1 also infects macrophages and causes formation of syncytia in culture, a characteristic similar to that of HIV (GARTNER et al. 1986).

In addition to the destruction of T cells, *ts*1 also disturbs the normal function of B cells. B cells cultured in vitro with *ts*1 showed marked proliferation leading to

Table 3. Comparison of immune and neurologic parameters in AIDS and *ts*1-mouse model

AIDS	*ts*1-Mouse model
1. Immune parameters	1. Immune parameters[a]
a) Reduction in total number and functions of T-lymphocytes in particular helper T cells	a) Reduction in total number of T-lymphocytes
b) Lymphopenia	b) Lymphopenia
c) Reduced lymphocyte blastogenesis	c) Reduced lymphocyte blastogenesis
d) Secondary infections	d) Secondary infections
e) Infection of macrophages and induction of syncytia	e) Infection of macrophages and induction of syncytia
f) Direct polyclonal activation of B-lymphocytes	f) Direct polyclonal activation of B-lymphocytes[b]
2. Neurologic parameters	2. Neurologic parameters
a) Vacuolar myelopathy in brain, brain stem, and gray and white matter of the spinal cord	a) Vacuolar myelopathy in brain, brain stem, and gray and white matter of the spinal cord
b) No inflammatory reaction	b) No inflammatory reaction
c) Few virus particles detected in CNS budding from cell membranes of neuron, glial, and capillary endothelial cells	c) Few virus particles detected in CNS budding from cell membranes of neuron, glial, and capillary endothelial cells

[a] WONG et al. (1989)
[b] G. PRASAD and P.K.Y. WONG (unpublished observation)

increased nonvirus-specific IgG and IgM synthesis, indicating polyclonal activation (G. PRASAD and P.K.Y. WONG, unpublished data). The ability of *ts*1 to cause direct polyclonal activation of B-lymphocytes is another characteristic that is similar to that of HIV (SCHNITTMAN et al. 1986). Preliminary in vivo studies indicate that *ts*1-infected mice showed hypergammaglobulinemia (G. PRASAD and P.K.Y. WONG, unpublished data). This phenomenon observed in vitro may be responsible for the polyclonal B-cell activation seen in *ts*1-infected mice.

It is clear from these studies that *ts*1 infection of BALB/c mice causes both immune and CNS system dysfunctions. It is interesting to note that *ts*1-infected mice develop a spectrum of symptoms which have many similarities to those found in HIV infections in humans. A comparison of the immune and neurologic parameters in human AIDS and the *ts*1-mouse model is presented in Table 3. Although the *ts*1-mouse model is evidently not a homolog of HIV-induced AIDS, it could serve as a valuable tool to investigate whether there is a common mechanism by which retroviruses impair the function of the immune system and interfere with neuronal function.

12 Molecular Mechanism of Pathogenesis Induced by the Paralytogenic Mutants

Previous studies (WONG et al. 1983) in this author's laboratory suggest that the ability of the paralytogenic *ts* mutants to induce hind limb paralysis is due to specific alteration(s) in the viral genome and is not a common characteristic among all the *ts* mutants of MoMuLV-TB. Since these mutants are derived from a known nonparalytogenic parent, the exact gene or gene sequence that confers on these mutants the ability to induce paralysis can be elucidated. Identification of the gene sequence(s) and function(s) responsible for the neurovirulence of the mutants should help to gain insight into the molecular mechanism of the pathogenesis of this neurologic disease. To pursue this goal, the subgenomic fragment(s) in the genome of *ts*1 which contain the critical mutations responsible for neurovirulence were identified using a panel of recombinant viruses constructed by exchanging homologous subgenomic fragments between genomes of molecularly cloned *ts*1 and either MoMuLV-TB or MoMuLV. Infectious virus obtained from these recombinant constructs upon transfection into NIH/3T3 cells were tested for their temperature sensitivity, ability to process $Pr80^{env}$ to gp70 and Prp15E, replication efficiency in the CNS, and ability to induce paralysis (YUEN et al. 1985a, b, 1986). Two domains in the genome of *ts*1 are found to be responsible for its paralytogenic ability. A 0.77-kbp *XbaI*-*Bam*HI *ts*1 fragment which encodes for the 5′ half of gp70 was found to be responsible for the temperature sensitivity and inability of *ts*1 to process $Pr80^{env}$. The other region, the 2.3-kbp *Bam*HI-*Pst*I fragment, which encompasses nearly two-thirds of the *env* gene, the entire LTR, and the 5′ noncoding region was found to be responsible for the enhanced neurotropism of *ts*1. Replacement of one or the other subgenomic fragment with the homologous fragment from either one of the wt viruses resulted in recombinant viruses which totally failed to induce paralysis or induce a greatly attenuated form of paresis in a very small percentage of the infected mice (YUEN et al. 1986). Using another recombinant construct (SZUREK et al. 1988), the region responsible for neurotropism was further delimited to the 1.1-kbp *Bam*HI-*Cla*I fragment which encodes the 3′ half of gp70 and almost all of p15E. These studies indicate that the neurovirulent determinants of *ts*1 are located entirely within the *env* gene.

A comparison of the nucleotide sequence of the entire *env* gene of *ts*1 with those of MoMuLV and MoMuLV-TB was carried out to pinpoint the mutations in the *ts*1 envelope gene which are responsible for its neurovirulence (SZUREK et al. 1988). This comparison showed that MoMuLV-TB and *ts*1 have 11 amino acid differences relative to MoMuLV in a region of the *env* gene bound by *Xba*I and *Cla*I. Of these 11 amino acid differences, 4 are unique to *ts*1. Two amino acids, Ile and Ala, are unique to the *ts*1 *Xba*I-*Bam*HI region, while the other two amino acids, Lys and Val, are unique to the *ts*1 *Bam*HI-*Cla*I domain (Fig. 3a). These results indicate that the unique Ile or Ala at the amino-terminal end of the envelope polyprotein is responsible for *ts*1 temperature sensitivity and inefficient processing of $Pr80^{env}$. The unique Lys at the carboxy end of gp70 or the unique Val at the amino-terminal end of p15E is responsible for *ts*1 neurotropism (SZUREK et al. 1988). In a very recent study, the unique amino acid at the amino-terminal end of p15E of *ts*1 was found to be not responsible for *ts*1's neurovirulence (SZUREK et al., in press). Thus, the neurovirulence determinants of *ts*1 are located entirely in the gene which encodes for the external envelope glycoprotein, gp70.

The mutation(s) within the *ts* Mo BA-1 virus genome which confers the virus the ability to induce neurodegenerative disease has not been determined (BILELLO et al. 1986). In the WM-E virus system, initial studies by OLDSTONE et al. (1980) suggested that both gp70 and p30 proteins of WM-E virus are necessary for induction of paralysis. Subsequent experiments by RASHEED et al. (1983) further implicated that gp70 of the WM-E virus is responsible for paralytogenesis. Analysis of the molecular determinants for paralysis induced by WM-E virus was later performed by DES GROSEILLERS et al. (1984), using chimeric viruses constructed by exchanging restriction fragments between Cas-Br-E and a nonparalytogenic amphotropic MuLV. These workers located the paralysis-inducing determinant of WM-E virus within a 3.9-kbp *Sal*I to *Cla*I restriction fragment which encodes for the 3′ end of *pol* and most of *env*. The nucleotide sequence of the 3.9-kbp fragment has been determined (RASSART et al. 1986) and the amino acid sequence deduced from the nucleotide sequence. When the amino sequence of the Cas-Br-E *env* protein was compared with the amino acid sequences of the *env* proteins of four different MuLVs namely MoMuLV, AKV, F-MuLV, and MCF-247, many differences were found (RASSART et al. 1986). Since a genetically closely related nonparalytogenic parent of Cas-Br-E is not available, the site(s) of the paralysis-inducing determinant could not be pinpointed.

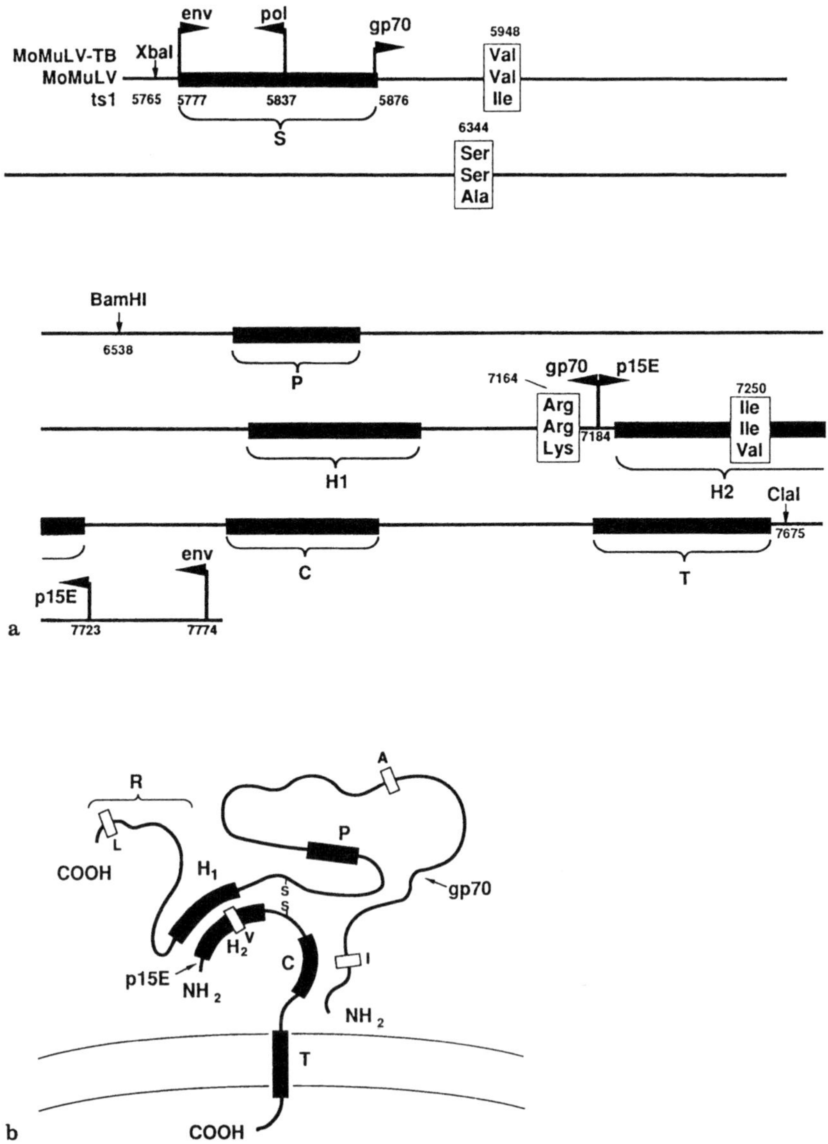

Fig. 3 a,b. Schematic representation of structural features of MuLV *env* protein. a Linear presentation of the *env* precursor indicating positions of different domains (*heavy solid lines*) within the *env* precursor and location of the four amino acid alterations (*box*) which are found in *ts*1 but not in MoMuLV and MoMuLV-TB. The amino acid sequences of *ts*1, MoMuLV, and MoMuLV-TB were deduced from the respective nucleotide sequences presented in Szurek et al. (1988). b Diagrammatic representation of the potential structure and orientation of gp70 and p15E relative to the lipid portion of the membrane. Positions of the amino acids unique to *ts*1 (*box*) and the different domains (*heavy solid lines*) are presented. *S* signal sequence; *P* proline-rich domain; H_1 gp70 hydrophobic domain; H_2 p15E hydrophobic domain; *C* highly conserved C-terminal *env* domain in p15E; *T* p15E transmembrane domain; *R* domain involves interaction with specific host cell receptor(s); *A* Ala; *I* Ile; *L* lys; *V* val

SZUREK et al. (1988) compared the unique amino acids encoded by the *ts*1 *env* gene with amino acids of five other MuLVs, namely MoMuLV, MoMuLV-TB, Cas-Br-E, AKV, and F-MuLV, aligned according to RASSART et al. (1986). In the *Xba*I-*Bam*HI fragment, which contains the 5′ half of gp70, there are two unique *ts*1 amino acids, Ile and Ala. At the *ts*1 Ile position all of the other MuLVs have a Val. The Ala of *ts*1 is replaced by Ser in four of the MuLVs, except for Cas-Br-E, which has a Thr. The *Bam*HI-*Cla*I fragment contains two unique *ts*1 amino acids, Lys and Val. At the Lys position, F-MuLV also has a Lys, but Cas-Br-E has a Pro, whereas the other three MuLVs have an Arg. The remaining amino acid unique to *ts*1 resides in the p15E hydrophobic domain but this amino acid has recently been shown not to be involved in paralytogenesis induction by *ts*1 (SZUREK, et al., in press).

It is interesting to note that changes in amino acid also occur in Cas-Br-E and F-MuLV at the same position as the unique *ts*1 amino acid, Lys, at the carboxy end of gp70, which is responsible for *ts*1's enhanced neurotropism. Although F-MuLV is nonparalytogenic, paralytogenic strains have been isolated by passaging the F-MuLV strain in rats (KAI and FURUTA 1984). One may hypothesize that the alteration in amino acid at this position may also render Cas-Br-E and F-MuLV more neurotropic.

Furthermore, of the two unique *ts*1 amino acids, Ile and Ala, recent data show that Ile is the one to be responsible for *ts*1's defect in processing of $Pr80^{env}$ to gp70 and Prp15E (SZUREK et al., in press). At this position Val is conserved in all the MuLVs. Since only *ts*1 has been reported to be defective in processing of $Pr80^{env}$ whereas the other MuLVs appear to be able to process the precursor envelope polyprotein, the change from Val to Ile is responsible for *ts*1's defect in processing $Pr80^{env}$ protein, which is in accord with the findings by SZUREK et al. (in press) mentioned above.

13 Discussion and Concluding Remarks

In the field of studies on virus-induced neurologic diseases, many excellent investigations have been performed on the biologic and pathologic aspects of the pathogenesis but very little has been reported on the molecular events leading to the neurologic disorders. This is largely due to: (a) the nature of the complex interactions between viruses and their hosts, (b) the cryptic nature of some of the etiologic agents, some of which are not amenable to molecular studies, (c) the long latent period in most of the chronic virus infections, and (d) the paucity of animal models that simulate human diseases.

During the past few years, however, studies on virus-induced neurologic disorders in animals using modern molecular and genetic approaches have begun to direct toward defining the molecular mechanism of viral pathogenesis. One of the most useful approaches in these studies is to compare the genetic makeup of closely related viral strains with a difference in neurovirulence (Atkins and Sheahan 1982; Dietzschold et al. 1985; Gonzalez-Scarano et al. 1985; Jackson et al. 1988; Janssen et al. 1984; Lustig et al. 1988; Seif et al. 1985; Spriggs et al. 1983; Stanley et al. 1985; Szurek et al. 1988; Wong et al. 1983, 1985; Yuen et al. 1985a, b, 1986).

The most crucial aspect of study on the pathogenesis of virus-induced neurologic disease is to understand the viral and host factors which contribute to the outcome of the disease. For a neurovirulent virus to inflict injury to the neural tissue, the virus must possess the ability to spread to the CNS. Once a virus has entered the CNS, the ability of the virus to interact with specific receptors on the plasma membrane of the cell plays a critical role in the subsequent events which lead to neurovirulence. The attachment of the viral envelope component to the receptor sets the stage for penetration of the virus into the cell either by fusion of viral membrane with host cell membrane (McClure et al. 1988; Stein et al. 1987) or by endocytosis (Mellman et al. 1986). Once inside the cell the virus will utilize the host cell's biosynthetic machinery to synthesize viral components. It is usually the synthesis and expression of the virus components in these cells that contribute to the pathogenesis of the disease. However, it is the adsorpton or attachment step which often plays a critical role in host specificity and tissue or cell tropism (for review see Johnson 1982 and Tyler 1988).

The differences in neurovirulence of *ts*1 and wt MoMuLV-TB could be attributed to two substitutions in the envelope gene of *ts*1 which result in two key pathologic features: (a) enhanced neurotropism and (b) neurocytotoxicity. Both enhanced neurotropism and neurocytotoxicity are required for the neurodegenerative disease (Yuen et al. 1986). The first feature involves viral recognition and interaction with specific receptors at the nerve cell surface. The second feature involves effect of the virus on the host cell as a result of viral replication (e.g., due to a viral-specific product).

To understand viral neurotropism, a study of the mechanism of virus attachment to neuronal and glial cells is necessary. However, in the MuLV system, the mechanism of virus attachment to host cell receptors has not been very well characterized at the molecular level. The most extensively studied animal virus-cell receptor interaction is that of the influenza virus sytem. A three-dimensional view of such a process has been reported very recently by Weis et al. (1988). In influenza virus the trimeric hemagglutinin molecule is responsible for cell recognition. The hemagglutinin molecule consists of an 85-Å-long stem which is attached to a large globular head. Four nonoverlapping epitopes that bind neutralizing antibodies have been mapped onto the surface of each monomer (Wiley et al. 1981). There is a small depression in the hemagglutinin surface between three of the four neutralizing epitopes. Amino acid residues lining this depression are highly conserved in comparison with the hypervariable antibody-binding sites. This surface feature has been shown to be the binding site to the cellular receptor (Weis et al. 1988).

Similar to influenza virus, MuLV adsorption involves the interaction of a small area of the gp70 protein with a specific host cell receptor protein (DeLarco and TODARO 1976). In ^{125}I-gp 70-binding inhibition studies, it has been shown that all ecotropic MuLVs belong to the same interference class and bind to one type of cell surface receptor (SARMA et al. 1967). The ecotropic cell surface receptor is different from the amphotropic receptor (GAZDAR et al. 1977). The host cell receptor for ecotropic MuLV has been isolated and partially purified (JOHNSON and ROSNER 1986). The number of receptors per cell vary from cell type to cell type (CHOPPIN et al. 1981). It has been reported that the ecotropic MuLV receptor is present in nerve cells (MOHAN and PAL 1982). However, all neurons do not contain the same receptors at their cell surface (KRISTENNSON and NORBY 1986; TYLER 1988). Possibly the paralytogenic MuLVs bind to a unique receptor on neuronal and/or glial cells which is not the same as the ecotropic MuLV receptor. Further studies are needed to resolve this problem.

The component of gp70 which interacts with cellular receptor resides in the carboxy end of the protein molecule (PINTER et al. 1982). A point mutation in the carboxy end of gp70 of *ts*1 results in a change in the amino acid residue from Arg to Lys (SZUREK et al. 1988). This mutation has been implicated to be associated with the neurotropism of *ts*1. This single amino acid substitution in *ts*1 may alter the conformation of the receptor-interacting component of gp70 at the carboxy end of the protein and change its ability to interact with the neuronal and/or glial cell receptor (Fig. 3B). Small genetic changes, including a single amino acid change in the host cell receptor-binding proteins for both rabies virus (DIETZSCHOLD et al. 1983; SEIF et al. 1985) and reovirus type 3 (BASSEL-DUBY et al. 1986; KAYE et al. 1986), have been found to account for marked differences in the neurovirulence of these viruses.

Neurotropism for motor neurons also appears to be a special property of the wild-mouse ecotropic viruses which is attributed to a specific affinity of the envelope glycoprotein for the surface receptor of neurons. This property as suggested by GARDNER (1985) must depend on a relatively small portion of the WM-E virus envelope. As mentioned earlier, it is interesting to note that, like *ts*1, a substitution of Arg residue near the carboxy end which is implicated to be responsible for *ts*1's neurotropism also occurs in the WM-E virus. Unless this is a coincidental event, the site for neurotropism of WM-E virus may also reside in this region of the envelope protein. Identification of the alteration of other neurotropic *ts* mutants of MoMuLV-TB in this region may further define the domain responsible for neurotropism.

The second key pathologic feature of *ts*1 which is responsible for its neurovirulence is intracellular accumulation of $Pr80^{env}$ polyprotein. Recent studies suggest that transport of newly made membrane proteins from the rough endoplasmic reticulum (RER) to the Golgi is a specific and highly regulated event (for review see LODISH 1988). These proteins undergo extensive modifications and folding while still in the RER. Unless proteins have achieved a proper configuration, they will not be transported to the Golgi apparatus. The substitution of amino acids, Val to Ile, in *ts*1 precursor envelope protein, $Pr80^{env}$, may prevent it from assuming the proper conformation, thereby hindering its transportation from RER to the Golgi complex to be processed by a cellular protease to gp70 and Prp15E (SZUREK et al. 1990). As a result they accumulate in the RER. It is plausible that accumulation of Pr80^{env} precursor protein cause dilatation of the endoplasmic reticulum and Golgi complex of the neuron, which has the most pronounced dilatation of RER and Golgi complex of all cells examined in the CNS (ZACHARY et al. 1986). Although this ultrastructure change in the neuron could be secondary to dysfunction of supporting glial cells, ZACHARY et al. (1986) considered that a primary change caused by neuronal infection with *ts*1 virus is more likely because experimental and spontaneous diseases with severe dysfunction of astrocytes are, in general, not associated with dilatation of neuronal RER or Golgi complex (ADACHI et al. 1973; ZACHARY and O'BRIEN 1985). Whether accumulation of precursor envelope proteins result in dilatation of the RER and Golgi complex of neurons in *ts*1-infected CNS remains to be resolved. The envelope precursor polyprotein could also accumulate at the cytoskeleton and cause degeneration of the nerve cells. Intracellular accumulation of glycoproteins has been identified by other investigators as being responsible for neurocytotoxicity in other neurologic disorders. In scrapie, kuru, and Creutzfedt-Jacob disease (CJD), it has been confirmed by several groups that agents which are mainly responsible for spongiform degeneration of the CNS are glycoproteins, and intracellular accumulation of this protein makes the neuronal cell membrane the primary target of the disease (PRUSINER and DEARMOND (1987). In neurons of Alzheimer's diseased brain tissue, there is a 68-d protein which is antigenically related to a smaller 59-d protein found in normal brain tissue (WOLOZIN et al. 1986). This larger protein seems to be present in higher amounts than the 59-d protein and is found in association with the characteristic neuritic plaques and neurofibrillary tangles in neuron of Alzheimer's brain.

In spite of the progress made in defining the pathology and genetic basis of paralytogenesis induced by the *ts*1 mutant in mice, gaps in our understanding on the mechanism involved in the pathogenesis of this disease remain. The following areas require further investigations: Does the relatively stereotypic symptomatology of degenerative diseases of the CNS induced by *ts*1 and other neurovirulent MuLVs result from the destruction of selective neuronal system? Is *ts*1's interaction with neuron and glial cells different from its interaction with other cell types? Is there any common factor(s) which makes T-lymphocyte and neuron more vulnerable to infection by *ts*1 virus? Is the cytopathic mechanism of *ts*1 in T-lymphocyte similar to that in neural cells? What role does the thymus play in the susceptibility of the infected mice to *ts*1? What are the early events in the course of the disease induced by *ts*1? What is the mechanism(s) of transmission of *ts*1 and other neurovirulent MuLVs to the CNS? What makes the substitution of Arg with a Lys residue at the carboxy terminal of gp70 enhance its interaction with the receptor of the neuron? Is the degeneration of motor neurons in *ts*1-infected mice due to a direct effect of *ts*1 infection of these cells or is this due to an indirect effect resulting from the infection of supportive cells (e.g., astrocyte)? Is the degeneration of the neuron due to the accumulation of Pr80^{env}? If so, does it require complete viral expression? Or does it only require the presence of increased amounts of unprocessed precursor envelope polyprotein? Where does the Pr80^{env}

accumulate in the cell? If the accumulation of unprocessed Pr80^{env} polyprotein is involved, how does this damage the function of nerve cells? Does it impair normal cell membrane transport functions that lead to cell degeneration, does it impair axonal transport resulting in axonal neuropathies, or does it cause dilatation of RER and Golgi complex resulting in vacuolization of the nerve cells? Since the induction of paralysis in the infected host by *ts*1 is host age dependent, is this due to an age-dependent restriction of virus replication in the infected CNS cells? Does productive virus replication in neurons and supportive cells in young mice affect neuronal cell functions differently relative to the effect the restricted virus replication may have in older nerve cells?

Further studies with this animal model together with others such as *ts* Mo BA-1 MuLV and the WM-E model should not only help us to understand the mechanism involved in the MuLV induction of neurologic disorder, it should also shed light on certain aspects of retrovirus replication such as the mechanism(s) involved in the processing of the envelope precursor polyprotein, factor(s) which control viral neurotropism, and the effect of the viral protein(s) on the host cell. Furthermore, information gained from these animal models may be useful for achieving a better understanding of human and other animal CNS diseases that are also characterized by spongiform changes or chronic neuronal degeneration. The increased understanding of retroviral pathogenesis at the genetic and molecular level that is based on an animal model should provide important new approaches for prevention and treatment of the diverse diseases caused by this unique group of retroviruses.

Acknowledgments. This chapter is dedicated to my wife, P.H. Yuen, whose encouragement and support is instrumental to the studies of *ts*1 and MoMuLV-TB presented in this review. I wish to thank my colleagues Benjamin Brooks, Judy Ball, George Stoica, Paul Szurek, Gaya Prasad, and Lezlie Coghlan for reviewing the manuscript and for their helpful comments. I also wish to thank Joyce Mayhugh for typing this manuscript. This work was supported by Public Health Service grants CA 45124 from the National Cancer Institute AI 28283 from National Institute of Allergy and Infections Disesess and a grant from Muscular Dystrophy Association.

References

Adachi M, Schneck L, Cara J, Volk BW (1973) Spongy degeneration of the central nervous system (van Bogaert and Bertrand type; Canavan's disease): a review. Hum Pàhol 4: 331–347

Anders KH, Guerra WF, Tomiyasu U, Verity MA, Vinters HW (1986) The neuropathology of AIDS. Am J Pathol 124: 537–558

Anderson RGW, Orci L (1987) A view of acidic intracellular compartments. J Cell Biol 106: 539–543

Andrews Jm, Gardner MB (1974) Lower motor neuron degeneration associated with type C RNA virus infection in mice: neuropathological features. J Neuropathol Exp Neurol 33: 285–307

Atkins GJ, Sheahan BJ (1982) Simliki Forest virus neurovirulence mutants have altered cytopathogenicity for central nervous system cells. Infect Immun 36: 333–341

Ball JK, Huh TY, McCarter JA (1964) On the statistical distribution of epidermal papillomata in mice. Br J Cancer 18: 120–123

Bassel-Duby R, Spriggs DR, Tyler KL, Fields BN (1986) Identification of attenuating mutations on the retrovirus type 3 Sl double stranded RNA segment using a rapid sequencing technique. J Virol 60: 64–67

Bedigian HG, Johnson DA, Jenkins NA, Copeland NG, Evans R (1984) Spontaneous and induced leukemias of myeloid origin in recombinant inbred BxH mice. J virol 51: 586–594

Bilello JA, Pitts OM, Hoffman PM (1986) Characterization of a progressive neurodegenerative disease induced by a temperature-sensitive Moloney murine leukemia virus infection. J Virol 59: 234–241

Brooks BR, Swarz JR, Johnson RT (1980) Spongiform polioencephalomyelopathy caused by a murine retrovirus. I. Pathogenesis of infection in newborn mice. Lab Invest 43: 480–486

Brooks Br, Gossage J, Johnson RT (1981) Age-dependent *in vitro* restriction of mouse neurotropic retrovirus replication in central nervous system-derived cells from susceptible Fv-1^{nn} mice. Trans Am Neurol Assoc 106: 238–241

Brooks Br, Feussner GK, Lust WD (1983) Spinal cord metabolic changes in murine retrovirus-induced motor neuron disease. Brain Res Bull 11: 681–686

Choppin J, Schaffar-Deshayes L, Debre P, Levy J (1981) Lymphoid cell surface receptor for Moloney leukemia virus envelope glycoprotein gp71. I. Binding characteristics. J Immunol 126: 2347–2351

Crispens CG (1978) In : Handbook on the laboratory mouse. Thomas, Springfield, pp. 139

DeLarco J, Todaro G (1976) Membrane receptors for the murine leukemia viruses: characterization using the purified viral envelope glycoprotein, gp71. Cell 8: 365–371

DesGroseillers L, Barrett M, Jolicoeur P (1984) Physical mapping of the paralysis-inducing determinant of a wild mouse ecotropic neurotropic retrovirus. J Virol 52: 356–363

DesGroseillers L, Rassart E, Robitaille Y, Jolicoeur P (1985) Retrovirus-induced spongiform encephalopathy: the 3'-end long terminal repeat containing viral sequences influence the incidence of the disease and the specificity of the neurological syndrome. Proc Natl Acad Sci USA 82: 8818–8822

Dietzschold B, Wunner WH, Wiktor TJ, Lopes AD, Lafon M, Smith CL, Koprowski H (1983) Differences in cell-to-cell spread of pathogenic and apathogenic rabies virus *in vivo* and *in vitro*. Proc Natl Acad Sci USA 80: 70–74

Dietzschold B, Wiktor TJ, Trojanowski JQ, MacFarland RI, Wunner WH, Torres-Anjel MJ, Koprowski H (1985) Differences in cell-to-cell spread of pathogenic and apathogenic rabies virus *in vivo* and *in vitro*. J Virol 56: 12–18

Gardner MB (1978) Type C viruses of wild mice: characterization and natural history of amphotropic, ecotropic, and xenotropic MuLV. In: Current topics in microbiology and immunology, vol 79. Springer, Berlin Heidelberg New York, pp 215–259

Gardner MB (1985) Retroviral spongiform polioencephalomyelopathy. Rev Infect Dis 7: 99–110

Gardner MB, Henderson BE, Officer JE, Rongey RW, Parker JC, Oliver C, Estes JD, Huebner RJ (1973) A spontaneous lower motor neuron disease apparently caused by indigenous type C RNA virus in wild mice. JNCI 51: 1243–1249

Gartner S, Markovitz P, Markovitz DM, Kaplan MH, Gallo RC, Popovic M (1986) The role of mononuclear phagocytes in HTLV-III/LAV infection. Science 233: 215–219

Gazdar AF, Oie H, Lalley P, Moss WW, Minna JD (1977) Identification of mouse chromosomes required for murine leukemia virus replication. Cell 11: 949–956

Goff S, Lobel LI (1987) Mutants of murine leukemia viruses and retroviral replication. Biochim Biophys Acta 907: 93–123

Gonzalez-Scarano F, Janssen RS, Najjar JA, Pbjecky N, Nathanson N (1985) An avirulent GIIglycoprotein variant of La Crosse bunyavirus with defective fusion function. J Virol 54: 757–763

Greenberger JS, Stephenson JR, Aaronson SA (1975) Temperature-sensitive mutants of murine leukemia virus. V. Impaired leukemogenic activity *in vivo*. Int J Cancer 15: 1009–1015

Haase AT (1986) Pathogenesis of lentivirus infections. Nature 322: 130–136

Hoffman PM, Ruscetti SK, Morse HC (1981) Pathogenesis of paralysis and lymphoma associated with a wild mouse retrovirus infection. Part I. Age and dose-related effects in susceptible laboratory mice. J Neuroimmunol 1: 275–285

Hoffman PM, Robbin DS, Morse HC (1984) Role of immunity in age-related resistance to paralysis after murine leukemia virus infection. J Virol 52: 734–738

Jackson AC, Moench TR, Trapp BD, Griffin DE (1988) Basis of neurovirulence in Sindbis virus encephalomyelitis of mice. Lab Invest 58: 503–509

Janssen R, Gonzalez-Scarano F, Nathanson N (1984) Mechanisms of bunyavirus virulence: comparative pathogenesis of a virulent strain of La Crosse and an avirulent strain of Tahyna virus. Lab Invest 50: 447–455

Johnson PA, Rosner MR (1986) Characterization of murine-specific leukemia virus receptor from L cells. J Virol 58: 900–908

Johnson RT (1982) Selective vulnerability of neural cells to viral infections. Adv Neurol 36: 331–337

Kai K, Furuta T (1984) Isolation of paralysis-inducing murine leukemia viruses from Friend virus passaged in rats. J Virol 50: 970–973

Kaye KM, Spriggs DR, Bassel-Duby R, Fields BN, Tyler KL (1986) Genetic basis for altered pathogenesis of an immune-selected antigenic variant of reovirus type 3 (Dearing). J Virol 59: 90–97

Kobayashi S, Katayama M, Satoh J, Suzuki K, Suzuki K (1988) The twitcher mouse. An alteration of the unmyelinated fibers in the CNS. Am J Pathol 131: 30–319

Koeing S, Gendelman HE, Orenstein JM, Dal Canto MC, Pezeshkpour GH, Yungbluth M, Janotta F, Aksamit A, Martin MA, Fauci AS (1986) Detection of AIDS virus in macrophages in brain tissue from AIDS patients with encephalopathy. Science 233: 1089–1093

Kristensson K, Norrby E (1986) Persistence of RNA viruses in the central nervous system. Annu Rev Microbiol 40: 159–184

Lampert PW, Gajdusek DC, Gibbs CJ (1972) Subacute spongiform virus encephalopathies, scrapie, kuru, and Creutzfeldt-Jakob disease (a review). Am J Pathol 68: 626–646

Lodish HF (1988) Transport of secretory and membrane glycoproteins from the rough endolasmic reticulum to the Golgi. J Biol Chem 263: 2107–2110

Lustig S, Jackson AC, Hahn CS, Griffin DE, Strauss EG, Strauss JH (1988) Molecular basis of Sindbis virus neurovirulence in mice. J Virol 62: 2329–2336.

McCarter JA, Ball JK, Frei JV (1977) Lower limb paralysis induced in mice by a temperature-sensitive mutant of Moloney leukemia virus. JNCI 59: 179–183

McClure MO, Marsh M, Weiss RA (1988) Human immunodeficiency virus infection of CD4-bearing cells occurs by a pH-independent mechanism. EMBO J 7: 513–518

Mellman I, Fuchs R, Helenius A (1986) Acidification of the endocytic and exocytic pathway. Annu Rev Biochem 55: 663–700

Mohan S, Pal BK (1982) Binding characteristics of wild mouse type C virus to mouse spinal cord and spleen cells. Infect Immun 37: 532–538

Moloney JB (1960) Biological studies on a lymphoid-leukemia virus extracted from sarcoma 371. Origin and introductory investigation. JNCI 24: 933–947

Navia BA, Cho E-S, Petito CK et al. (1986) The AIDS dementia complex: II. Neuropathology. Ann Neurol 19: 525–535

Officer JE, Tecson N, Fontanilma E, Rongey RW, Estes JD, Gardner MB (1973) Isolation of a neurotropic type-C virus. Science 181: 945–947

Oie HK, Gazdar AF, Lalley PA, Russell EK, Minna JD, DeLarco J, Todaro GJ, Francke U (1978) Mouse chromosome 5 codes for ecotropic murine leukemia virus cell-surface receptor. Nature 274: 60–62

Oldstone MBA, Lampert PW, Lee S, Dixon FJ (1977) Pathogenesis of the slow disease of the central nervous system associate with WM-1504E virus. Am J pathol 88: 193–206

Oldstone MBA, Jensen F, Dixon FJ, Lampert PW (1980) Pathogenesis of the slow disease of the nervous system associated with wild mouse virus. II. Role of virus and Host gene products. Virology 107: 180–193

Oldstone MBA, Jensen F, Elder J, Dixon FJ, Lampert PW (1983) Pathogenesis of the slow disease of the central nervous system associated with wild mouse virus. III. Role of input virus and MCF recombinants in disease. Virology 128: 154–165

Osame M, Usuku K, Izumo S et al. (1986) HTLV-I associated myelopathy. A new clinical entity. Lancet i: 1031–1032

Pahwa R, Good RG, Pahwa S (1987) Prematurity, hypogamma-glubulinemia, and neuropathology with human immunodeficiency virus (HIV) infection. Proc Natl Acad Sci USA 84: 3826–3820

Pal BK, Mohan S, Nimo R, Gardner MB (1983) Wild mouse retrovirus-induced neurogenic paralysis in laboratory mice. 1. Virus replication and expression in central nervous system. Arch Virol 77: 239–247

Prasad G, Stoica G, Wong PKY (1989) The role of the thymus in the pathogenesis of hind-limb paralysis induced by *ts*1, a mutant of Moloney murine leukemia virus-TB. Virology 169: 332–340

Petito CK, Navia BA, Cho E-S, Jordan BD, George DC, Price RW (1985) Vacuolar Myelopathy

pathologically resembling subacute combined degeneration in patients with the acquired immunodeficiency syndrome. N Engl J Med 312: 874–879
Pincus T, Rowe WP, Lilly F (1971) A major genetic locus affecting resistance to infection with murine leukemia viruses. II. Apparent identity to a major locus described for resistance to Friend murine leukemia virus. J Exp Med 133: 1234–1241
Pinter A, Honnen WJ, Tung J-S, O'Donnell PV, Hammerling U (1982) Structural domains of endogenous murine leukemia virus gp70s containing specific antigenic determinants defined by monoclonal antibodies. Virology 116: 499–516
Pitts OM, Powers JM, Bilello JA, Hoffman PM (1987) Ultrastructural changes associated with retroviral replication in central nervous system capillary endothelial cells. Lab Invest 56: 401–409
Prusiner SB, DeArmond SJ (1987) Prions causing nervous system degeneration. Lab Invest 56: 349–363
Rasheed S, Gardner MB, Lai MMC (1983) Isolation and characterization of new ecotropic murine leukemia viruses after passage of an amphotropic virus in NIH Swiss mice. Virology 130: 439–451
Rassart E, Nelbach L, Jolicoeur P (1986) Cas-Br-E murine leukemia virus: sequencing of the paralytogenic regions of its genome and derivation of specific probes to study its origin and the structure of its recombinant genomes in leukemic tissues. J Virol 60: 910–919
Reddy EP, Dunn CY, Aaronson SA (1980) Different lymphoid target cells for transformation by replication—competent Moloney and Rauscher mouse leukemia viruses. Cell 19: 663–669
Robey WG, Dekaban GA, Ball JK, Poore CM, Fischinger PJ (1985) Thymotropic envelope gene recombinants of Moloney leukemia virus have highly conserved envelope structures. Virology 142: 183–196
Roman GC (1987) Retrovirus-associated myelopathies. Arch Neurol 44: 659–663
Rude R, Gallick GE, Wong PKY (1980) A fast replica plating technique for the isolation of post-integration mutants of the Moloney strain of murine leukaemia virus. J Gen Virol 49: 367–374
Sarma PS, Cheong MP, Hartley JW, Huebner RJ (1967) A viral interference test for mouse leukemia viruses. Virology 33: 180–184
Schnittman SM, Lane HC, Higgins SE, Folks T, Fauci AS (1986) Direct polyclonal activation of human B lymphocytes by the acquired immune deficiency syndrome virus. Science 233: 1084–1086
Seif I, Coulon P, Rollin PE, Flamand A (1985) Rabies Virulence: effect on pathogenicity and sequence characterization of rabies virus mutations affecting antigenic site III of the glycoprotein. J Virol 53: 926–934
Smith JE, Brooks BR, Dolan K (1988) Molecularly cloned neurotropic retrovirus induced vacuolar myelopathy: dynamics of vacuolar formtions. (to be published)
Spriggs DR, Bronson RT, Fields BN (1983) Hemagglutinin variants of retrovirus type 3 have altered central nervous system tropism. Science 220: 505–507
Stanley J, Cooper SJ, Griffin DE (1985) Alphavirus neurovirulence: monoclonal antibodies discriminating wild-type from neuroadapted Sindbis virus. J Virol 56: 110–119
Stansly PG (1965) Non-oncogenic infectious agents associated with experimental tumors. Prog Exp Tumor Res 7: 224–258
Stein BS, Gowda SD, Lifson JD, Penhallow RC, Bensch KG, Engleman EG (1987) pH Independent HIV entry into CD4-positive T cells via virus envelope fusion to the plasma membrane. Cell 49: 659–668
Swarz JR, Brooks BR, Johnson RT (1981) Spongiform polioencephalomyelopathy caused by a murine retrovirus. II. Ultrastructural localization of virus replication and spongiform changes in the central nervous system. Neuropathol Appl Neurol 7: 365–380
Szurek PF, Yuen PH, Jerzy R, Wong PKY (1988) Identification of point mutations in the envelope gene of Moloney murine leukemia virus TB temperature-sensitive paralytogenic mutant *ts*1: molecular determinants for neurovirulence. J Virol 62: 357–360
Szurek PF, Yuen PH, Ball JK, Wong PKY (1990) A Val-to-Ile substitution in the envelope precursor polyprotein, gPr80^{env}, is responsible for the temperature sensitivity, in efficient processing of gPr80^{env}, and neurovirulence of *ts*1, a mutant of Moloney murine leukemia Virus-TB. J Virol in press
Tyler KL (1988) Host and viral genetic factors which influence viral neurotropism. In: Rosenberg RN, Harding AE (eds) Molecular biology of neurological disease. Butterworths, London, pp 64–65
Uzman LL, Rumley MK (1958) Changes in the composition of the developing mouse brain during early myelination. J Neurochem 3: 170–184
Weis W, Brown JH, Cusack S, Paulson JC, Skehel JJ, Wiley DC (1988) Structure of the influenza virus haemagglutinin complexed with its receptor, sialic acid. Nature 333: 426–431

Wiley DC, Wilson IA, Skehel JJ (1981) Structural identification of the antibody-binding sites of Hong Kong influenza haemagglutinin and their involvement in antigenic variation. Nature 289: 373–378

Wolozin L, Pruchnicki A, Dickson DW, Davies P (1986) A neuronal antigen in the brains of Alzheimer patients. Science 231: 648–650

Wong PKY, Gallick GE (1978) Preliminary characterization of the temperature-sensitive mutant of murine leukemia virus which produces defective particles at the restrictive temperature. J Virol 25: 187–192

Wong PKY, McCarter JA (1974) Studies of two temperature-sensitive mutants of Moloney murine leukemia virus. Virology 58–396–408

Wong PKY, Russ LJ, McCarter JA (1973) Rapid, selective procedure for isolation of spontaneous temperature-sensitive mutants of Moloney leukemia virus. Virology 51: 424–431

Wong PKY, Soong MM, MacLeod R, Gallick GE, Yuen PH (1983) A group of temperature-sensitive mutants of Moloney leukemia virus which is defective in cleavage of *env* precursor polypeptide in infected cells also induces hind limb paralysis in newborn CFW/D mice. Virology 125: 513–518

Wong PKY, Knupp C, Yuen PH, Soong MM, Zachary JF, Tompkins WAF (1985) *ts*1, a paralytogenic mutant of Moloney murine leukemia virus TB, has an enhanced ability to replicate in the central nervous system and primary nerve cell culture. J Virol 55: 760–766

Wong PKY, Prasad G, Hansen J, Yuen PH (1989) *ts*1, a mutant of Moloney murine leukemia virus-TB, causes both immunodeficiency and neurologic disorders in BALB/c mice. Virology 170: 450–459

Yuen PH, Szurek PF (1989) The reduced virulence of the thymotropic Moloney murine leukemia virus derivative MoMuLV-TB is mapped to II mutations within the U3 region of the LTR. J Virol 63: 471–480

Yuen PH, Wong PKY (1977) Electron microscopy characterization of defectiveness of a temperature-sensitive mutant of MoMuLV restricted in assembly. J Virol 24: 222–230

Yuen PH, Malehorn D, Knupp C, Wong PKY (1985a) A 1.6-kilobase pair fragment in the genome of the *ts*1 mutant of Moloney murine leukemia virus TB that is associated with temperature sensitivity, nonprocessing of $Pr80^{env}$, and paralytogenesis. J Virol 54: 364–373

Yuen PH, Malehorn D, Nau C, Soong MM, Wong PKY (1985b) Molecular cloning of two paralytogenic, temperature-sensitive mutants, *ts*1 and *ts*7, and the parently wild-type Moloney murine leukemia virus. J Virol 54: 178–185

Yuen PH, Tzeng E, Knupp C, Wong PKY (1986) The neurovirulent determinants of *ts*1, a paralytogenic mutant of Moloney murine leukemia virus TB, are localized in at least two functionally distinct regions in the genome. J Virol 59: 59–65

Zachary JF, O'Brien DP (1985) Spongy degeneration of the central nervous system in two canine littermates. Vet Pathol 22: 561–571

Zachary JF, Knupp CJ, Wong PKY (1986) Non-inflammatory spongiform polioencephalomyelopathy caused by a neurotropic temperature-sensitive mutant of Moloney murine leukemia virus TB. Am J Pathol 124: 457–468

Caprine Arthritis Encephalitis Lentivirus Transmission and Disease

T. C. McGuire, K. I. O'Rourke, D. P. Knowles, and W. P. Cheevers

1 Introduction

Lentiviruses occur in several species of animals and those that share DNA sequence homology with human immunodeficiency virus type 1 (HIV-1) include equine infectious anemia virus (EIAV) of horses, maedi/visna virus of sheep, and caprine arthritis encephalitis virus (CAEV) of goats (Chiu et al. 1985; Sonigo et al. 1985; Pyper et al. 1986). Visna virus and related lentiviruses of sheep are distinguished from goat CAEV isolates by nucleic acid hybridization, DNA sequence, and immunologic reactivity (reviewed in Cheevers and McGuire 1988); however, the viruses can reciprocally infect sheep or goats (Banks et al. 1983).

Initial reports of the CAE syndrome described a high incidence of encephalitis in young kids from a dairy goat herd (Cork et al. 1974a, b). The only significant

Department of Veterinary Microbiology and Pathology, College of Veterinary Medicine, Washington State University, Pullman, WA 99164, USA

Current Topics in Microbiology and Immunology, Vol. 160

disease noted in the clinical history of mature goats from the herd was arthritis (Cork et al. 1974a). Retroviruses were subsequently isolated from the joint of an arthritic goat (Crawford et al. 1980a; Cheevers et al. 1981) and the thymus from a kid inoculated with brain tissue from a natural case of encephalitis (Cork and Narayan 1980; Narayan et al. 1980). Experimental inoculation of these viruses induced inflammatory lesions in both brain and spinal cord (Adams et al. 1980b; Cork and Narayan 1980). Significant lesions caused by CAEV also occurred in the mammary gland (Cork and Narayan 1980; Cheevers et al. 1988) and in the lung (Ellis et al. 1983a).

Caprine arthritis encephalitis virus infection is an important disease of goats because of the high prevalence of both infection and disease. Infected goats remain seropositive for as long as they have been studied (Crawford and Adams 1981; Cheevers et al. 1988) and virus can be isolated from at least 70% of goats infected for several years. These data, taken together with data for persistence of other lentiviruses, suggest lifelong CAEV persistence. Prevalence, determined by serum antibody to CAEV antigens, was 81% in goats surveyed in the United States (Crawford and Adams 1981) and greater than 65% in some European countries (Adams et al. 1984). Estimates of the prevalence of clinical disease in infected goats range from 20% to 30% (Crawford and Adams 1981; Woodard et al. 1982; East et al. 1987).

There are several features of CAEV infection that merit attention. The virus causes natural disease in goats with virus persistence. Clinical signs occur relatively rapidly (within months) after CAEV infection, and the occurrence and severity of lesions are controlled by both the genotype of the host and the genotype of the virus isolate. A recent review of CAEV structure and virus-host cell interaction is available (Cheevers and McGuire 1988). The purpose of this review is to compile experimental CAEV transmission studies and suggest the natural mode of transmission, to compare encephalitis, arthritis, pneumonia, and mastitis in natural outbreaks and experimental transmission studies, and to discuss the variables associated with the host and with the virus which influence the outcome of infection.

2 Natural Mode of CAEV Transmission and Control

A high percentage (78%) of goat kids born to CAEV-infected dams acquired CAEV infection (Adams et al. 1983). The primary explanation for the high transmission rate from dam to kid was oral infection by ingesting CAEV in colostrum and milk. Evidence for this mode of transmission included several observations: (a) cesarean-derived kids from CAEV-infected dams that were isolated from the dam and fed CAEV-free cow or goat milk had a very low percentage (5%) of CAEV infection (Crawford et al. 1980a; Adams et al. 1980a, 1983); (b) goat kids removed from their CAEV-infected dams at birth, deprived of colostrum or milk, isolated, and fed cow milk had a very low percentage (4%–10%) of CAEV infection (ADAMS et al. 1983; ELLIS et al. 1983b); (c) goat kids born to CAEV-free dams and allowed to suckle colostrum and milk did not develop CAEV infection (ADAMS et al. 1983); (d) CAEV was isolated from milk cells from CAEV-infected dams by cocultivation with susceptible cells in culture (ADAMS et al. 1983; ELLIS et al. 1983b), and CAEV was isolated from cell-free milk from CAEV infected dams in 7 of 27 cases (ADAMS et al. 1983); (e) kids fed during the first 5 days of life were infected with milk pooled from CAEV-infected dams which contained CAEV-infected cells (ADAMS et al. 1983). Also, colostrum from CAEV-infected dams fed to CAEV-free kids for 7 days after birth caused CAEV infection in the kids (ELLIS et al. 1986).

Whether infrequent intrauterine CAEV infection of kids occurs or not is unclear. The small number of transmissions (2 of 37) that occurred in cesarean-derived kids that were isolated and fed cow milk could have been due to intrauterine infection or to other routes (ADAMS et al. 1983). Intrauterine infection occurs infrequently with ovine progressive pneumonia virus, a United States isolate of maedi/visna virus (CROSS et al. 1975; CUTLIP et al. 1981). In one of these studies, virus was isolated from 1 of 17 lamb fetuses from seropositive dams (CUTLIP et al. 1981). Natural transmission of maedi/visna virus in sheep by the intrauterine route was very infrequent (HOUWERS et al. 1983). In one study, 415 lambs were separated from the dam at birth, deprived of sheep colostrum, and reared in isolation from the herd, which had 63% seropositive dams, and only 2 of these lambs developed maedi/visna infection (HOUWERS et al. 1983).

Therefore, control programs for CAEV and for related sheep lentiviruses are based on preventing transmission from infected dams to the offspring by isolation of the offspring at birth and depriving the offspring of colostrum or milk from infected dams (ADAMS et al. 1983; ELLIS et al. 1983b; HOUWERS et al. 1983; MACDIARMID 1984; ROBINSON and ELLIS 1986). Isolation of offspring from infected animals is required because transmission can occur by contact both during and after the perinatal period. The evidence for contact transmission is from both experimental studies and evaluation of natural outbreaks: (a) goat kids born to CAEV-infected dams and prevented from suckling their dams, but licked by the dam, had a 17% prevalence of infection (ADAMS et al. 1983); (b) continuous direct exposure of five CAEV-free goats to CAEV-infected goats under pasture conditions resulted in the infection of two goats between 9 and 22 months of exposure (ADAMS et al. 1983); (c) continuous direct exposure of 15 seronegative females to seropositive females under dairy management conditions resulted in the CAEV-infection of nine females during a 10-month period (ADAMS et al. 1983); (d) introduction of apparently infected goats into a dairy herd resulted in a CAEV outbreak (WOODARD et al. 1982); (e) introduction of CAEV-infected goats into a country apparently free of CAEV resulted in the infection of native goats which had direct contact with the introduced goats (ADAMS et al. 1984); (f) no transmission was found between CAEV-infected and noninfected lactating female goats when maintained in an open barn, but having no physical contact (ADAMS et al. 1983); (g) prevalence of antibodies to ovine progressive pneumonia virus increased with age in range sheep, even after 7 years of age (GATES et al. 1978).

The mechanisms of contact transmission are unknown, but may involve the ingestion of milk or other secretions from infected animals. Oral transmission is not limited to young animals since eight weaned goats were infected by ingesting a single dose of cell-free tissue culture supernatant containing 1×10^6 50% tissue culture infective doses ($TCID_{50}$) of CAEV (T.C. MCGUIRE, unpublished results).

In conclusion, the primary mode of transmission of CAEV is by kids ingesting colostrum or milk from their infected dams. Transmission occurs by direct contact in all age groups and this contact transmission may also be oral ingestion of secretions from infected animals. One likely explanation for the very high prevalence of CAEV infection in goats in the United States and several other countries where dairy goats are common was the practice of pooling colostrum from several dams and feeding it to several kids (ADAMS et al. 1984). This explanation is supported by the low prevalence (0.1%) of CAEV antibody in goats raised for fiber in Australia (SURMAN et al. 1987).

3 Lesions Caused by CAEV Infection

3.1 Leukoencephalomyelitis

3.1.1 Natural Leukoencephalomyelitis

Caprine arthritis encephalitis virus was discovered through studies of a natural outbreak of encephalitis in a dairy goat herd (CORK et al. 1974a). Appearance of encephalitis was followed in kids born to yearling dams on a goat dairy farm (CORK et al. 1974a). Yearling dams born on the farm were used to replace aging dams. In the 1st year of study, 19 kids were born to 10 yearling dams, and 16% of the kids developed encephalitis. In the next year, 14 yearling dams had 21 kids, and 19% developed encephalitis. The only significant finding in the rest of the herd was clinical arthritis in adults (CORK et al. 1974a).

The onset of nervous system signs in the original description (CORK et al. 1974a) and in subsequent studies of natural disease (O'SULLIVAN et al. 1978; NORMAN and SMITH 1983) was primarily in kids, 1–5 months of age. Sporadic neurologic disease was reported in adult goats and a lentivirus was isolated (SUNDQUIST et al. 1981). Signs of brain involvement in kids included circling, facial twitching, head tilt, tremors, and torticollis (NORMAN and SMITH 1983; NARAYAN and CORK 1985). The most dramatic signs reported were attributed to spinal cord involvement and included marked weakness in one or more hind limbs which progressed to para- or tetraparesis (CORK et al. 1974a; NORMAN and SMITH 1983). Evaluation of the CSF revealed increases in leukocytes (mostly lymphocytes) and protein (CORK et al. 1974a; NORMAN and SMITH 1983).

Caprine arthritis encephalitis virus-caused inflammatory lesions were in the brain and spinal cord (CORK et al. 1974a, b; NORMAN and SMITH 1983). The leukoencephalomyelitis consisted of focal accumulations of macrophages, lymphocytes, and plasma cells in perivascular areas and infiltrating into the white matter. Demyelination occurred (CORK and DAVIS 1975) and in severe cases there were large areas of malacia. Demyelinating and inflammatory lesions coexisted in some animals (NARAYAN and CORK 1985). In the spinal cord, lesions occasionally progressed to cavitation and extended into adjacent gray matter (CORK et al. 1974b). In all cases reported, there was a mononuclear cell inflammatory component.

3.1.2 Experimental Leukoencephalomyelitis

Variability in the development of CNS disease has been reported following experimental transmissions of CAEV. Combined intracerebral and intraperitoneal inoculation of brain filtrate caused clinical signs of CNS disease in two of five goats and CNS lesions in four of five goats examined 23–37 days after inoculation (CORK et al. 1974a). The lesions were those of leukoencephalomyelitis described in natural disease, although less severe. Inoculation of tissue culture virus by combined intracerebral, intraarticular, and intraperitoneal routes resulted in leukoencephalomyelitis similar to that in natural disease (CORK and NARAYAN 1980). The CNS lesions were induced in both newborns and young adult goats. However, these lesions radiated from the inoculation site localizing around the ventricular system, vessels, and fiber tracts. Another CAEV isolate also induced CNS lesions after intracerebral inoculation (SMITH et al. 1981).

When CAEV was inoculated by combined intravenous and intraarticular routes (CRAWFORD et al. 1980a; ADAMS et al. 1980b; ELLIS et al. 1983a), mild perivascular mononuclear cell infiltrates were the only CNS lesions reported. None of these studies reported CNS signs.

A recent experiment evaluated CAEV-induced disease 3 years after oral infection of newborn kids with two different biologically cloned isolates of CAEV (CHEEVERS et al. 1988). One isolate (CAEV 63) was isolated from explanted carpal synovium from an adult goat with a natural case of chronic arthritis (CRAWFORD et al. 1980a), while the other (CAEV Co) was isolated from explanted thymus from a goat inoculated with brain suspension from a goat with a natural case of encephalitis (NARAYAN et al. 1980). Both isolates were biologically cloned by terminal dilutions and propagated on fetal caprine synovial cell cultures. Seventeen newborn CAEV-free goat kids were fed once/day for the first 5 days of life with inoculum containing 9.55×10^5 $TCID_{50}$ of CAEV 63, and 18 goats were similarly fed 9.55×10^5 $TCID_{50}$ of CAEV Co (CHEEVERS et al. 1988). At 3 years after infection, lesions were confined to the joint and mammary gland, except for histologic evidence of mild meningitis in 2 of the 35 goats. Spinal cord was not examined; however, clinical indications of myelitis were not detected.

The reasons for the difficulty in experimentally reproducing the leukoencephalomyelitis noted in natural cases of CAEV without intracerebral inoculation of brain tissue containing CAEV are not known. There are several interesting possibilities which are related to the virus. (a) Genetic changes in the virus may cause the variability noted in organ specificity and the severity of lesions. For example, infectious mutants of HIV-1 with changes in the 3′ region had reduced cytopathic effects in vitro (FISHER et al. 1986). Also, the virulence of two genetically distinct CAEV isolates was different (CHEEVERS et al. 1988). (b) In the oral inoculation study (CHEEVERS et al. 1988), the CAEV isolates used were biologically cloned and propagated in fetal goat synovial membrane cells, which may have selected against neurotropic variants. Genetically different isolates of HIV-1 with distinct cellular tropisms were isolated from the brain tissues of a patient with AIDS encephalopathy (KOYANAGI et al. 1987). (c) Natural CAEV infection of kids with colostrum and milk by the oral route involves infected milk cells and cell-free virus which may be associated with antibody. Whether virus in these forms influences the course of infection is unknown. (d) Clinical signs noted in natural cases of CAEV were severe and correlated with very marked brain and spinal cord lesions (CORK et al. 1974a; NORMAN and SMITH 1983). Morphologic changes noted in AIDS encephalopathy and dementia are usually very subtle (PRICE et al. 1988), and the initial clinical features involve changes in cognitive and behavior functions with some motor dysfunction. Such morphologic changes were not apparent in experimental CAEV-infected goat brains; however, cognitive and behavior changes would be difficult to interpret.

3.2 Arthritis

3.2.1 Natural Arthritis

Arthritis was noted early in studies of CAEV infections (CORK et al. 1974a). In contrast to CNS lesions, arthritis occurred predominantly in adults (CRAWFORD et al. 1980a, b). Features of the natural disease in adult goats included chronic arthritis with swollen peripheral joints, particularly the carpus. The pathologic changes also involved the tendon sheaths, bursae, and adjacent tissues (CRAWFORD et al. 1980b; WOODARD et al. 1982). Lesions in synovial-lined structures included synovial cell proliferation, subsynovial mononuclear cell infiltration, vascular injury and leakage, fibrin coating of villi, villi proliferation, fibrin concretions, villi necrosis, and mineralization. In goats with chronic arthritis, there were also emaciation, rough and long hair coat, lameness and occasional CNS signs (CRAWFORD et al. 1980b). In one natural CAEV outbreak, the clinical signs were carpal hygroma and stiffness, progressing to a debilitating lameness in 30% of the affected goats (WOODARD et al. 1982). In this outbreak, both kids and adults had arthritis, but CNS signs were not a prominent factor. Isolation of CAEV was made from tissues taken at necropsy from 24 of 27 goats with arthritis (SURMAN et al. 1987).

3.2.2 Experimental Arthritis

Arthritis was readily reproduced by intraarticular inoculation of CAEV (CRAWFORD et al. 1980a; CORK and NARAYAN 1980; ADAMS et al. 1980b; ELLIS et al. 1983a). In experiments where combined intravenous and intraarticular injection of one carpus was done, arthritis occurred in both the inoculated and the uninoculated carpi. Experimental arthritis began with synovial cell hyperplasia and perivascular mononuclear cell infiltration (ADAMS et al. 1980b). This progressed to severe synovial cell hyperplasia, subsynovial mononuclear cell infiltration, and villi hypertrophy. Eventually the villi resembled lymph node tissue with formation of lymphoid follicles. Vascular leakage, fibrin concretions, and necrosis were apparent, and clinical signs of arthritis included swollen joints and lameness.

A positive correlation occurred between CAEV expression in the synovial cavity and development of clinically detectable arthritis (KLEVJER-ANDERSON et al. 1984). Ten goat kids were inoculated with biologically cloned CAEV 63 intraarticularly and intravenously and examined periodically for 18 months. Virus was isolated from cell-free synovial fluid from all 10 goats at 1 month postinfection (p.i.), from 6/10 at 6 months p.i., from 4/10 at 12 months p.i., and from 3/10 at 18 months p.i. Joint swelling had a positive correlation with virus isolation from the joint when both cell-free virus and virus from synovial fluid cells were considered. Also, the synovial fluid contained 13- to 15-fold more cells than in synovial fluid from mock-infected control goats.

Evaluation of the immune response demonstrated that serum antibody titers to CAEV antigens had a peak at 45–80 days p.i. and declined to an easily detectable plateau until the termination of the experiment at 280 days p.i. (ADAMS et al. 1980a). The in vitro proliferation of peripheral blood mononuclear cells from CAEV-infected goats to CAEV antigens was detectable at 15 days p.i. and continued to increase until the end of the experiment at 271 days p.i. (ADAMS et al. 1980a). In the synovial fluid collected 6 months p.i., there was a 2- to 5.3-fold increase in polyclonal immunoglobulin G1 over the serum concentrations (JOHNSON et al. 1983a). This synovial fluid immunoglobulin G1 concentration was increased relative to mock-inoculated control goats at 38 months p.i. Evaluation of the antibody response to CAEV antigens in synovial fluid by western blotting and immunoprecipitation demonstrated high antibody titers to gp135 and gp90 in contrast to antibody titers to p28 (JOHNSON et al. 1983b). In another study of CAEV arthritis in 35 goats, the amount of anti-gp135 antibody in synovial fluid at 3 years p.i. reflected the severity of arthritis in all 70 carpal joints (D.P. KNOWLES, unpublished observations).

In summary, immune reactants including cells and immunoglobulin were present in joints with CAEV arthritis, a large amount of synovial fluid antibody was specific for CAEV antigens, and cell-free or cell-associated virus could be isolated from synovial fluid over several months. Therefore, it was postulated that arthritis resulted from inflammation caused by immune responses to CAEV. Support for this assumption came from unsuccessful attempts to induce immunity to CAEV by vaccination with inactivated CAEV (McGuire et al. 1986). CAEV-vaccinated goats developed more severe arthritis than did non-vaccinated goats after challenge inoculation with CAEV. In another experiment, goats with persistent CAEV-infection developed acute arthritis after inoculation of CAEV (McGuire et al. 1986).

The cell type infected by CAEV, both in vitro (Adams et al. 1980b) and in vivo, appears to be macrophages (Klevjer-Anderson and Anderson 1982; Anderson et al. 1983a; Narayan et al. 1983; Gendelman et al. 1985). The inflammation initiated by specific immune responses to CAEV may be augmented by effects of CAEV on macrophage function (DeMartini et al. 1983; Banks et al. 1987).

3.3 Pneumonia

The original description of natural CAE in kids in the United States described a mild interstitial pneumonia in which the interalveolar septae were infiltrated with mononuclear cells (Cork et al. 1974a). Similar findings were noted in a natural outbreak involving primarily adult goats (Woodard et al. 1982). In Australia, a progressive interstitial pneumonia was described in seven goats in which the pulmonary lesions were severe (Sims et al. 1983). All lobes of the lungs were consolidated, and mononuclear inflammatory cells accumulated in alveolar septae, and peribronchial and perivascular areas. There were also accumulations of eosinophilic material in the alveolar spaces. A virus similar to CAEV was isolated from one goat. Later, 50 Australian dairy goats that were seropositive to CAEV were examined postmortem (Ellis et al. 1988). Twenty-one had chronic interstitial pneumonia of caudal or cranioventral lobes with histologic changes similiar to those previously described. CAEV was isolated from the lungs in all 12 of the attempts made, although isolation of CAEV from the lungs of 3 goats with arthritis without pneumonia precluded drawing a cause and effect relationship.

None of the experimental transmissions with CAEV induced significant lung lesions in recipient animals (Adams et al. 1980b; Ellis et al. 1983a; Cheevers et al. 1988). The explanations for the inability to induce the lung lesions seen in natural cases with CAEV may be the same as those listed for encephalitis (see Sect. 3.1.2). Similar results were noted for ovine progressive pneumonia, even with intra-thoracic inoculation (Oliver et al. 1981). However, lymphoid interstitial pneumonia was rapidly induced in neonatal lambs with an ovine lentivirus isolate given intratracheally (Lairmore et al. 1986).

3.4 Mastitis

Descriptions of mastitis caused by CAEV were made in transmission studies with CAEV inoculated by a combination of intraarticular, intravenous, and intra-cerebral routes (Cork and Narayan 1980). Also, mastitis was a prominent feature of disease in goats 3 years after oral infection with two different biologically cloned CAEV isolates (CHEEVERS et al. 1988). In the latter study, all six females infected with CAEV 63 and seven of nine females infected with CAEV Co had mastitis. Microscopic lesions in the mammary gland were inflammatory and included infiltration of mononuclear cells into the periductal stroma. These cells included lymphocytes, macrophages, and plasma cells. The cellular infiltrates replaced normal structures, necrotic areas occurred, and in some cases the infiltrates resembled lymph node germinal centers. The frequent presence of CAEV mastitis probably results in cell-associated and cell-free CAEV in colostrum and milk, accounting for much of the natural transmission of CAEV to neonatal goat kids (ADAMS et al. 1983; ELLIS et al. 1983b; ZWAHLEN et al. 1983).

4 Host Genetic Factors and Disease Expression

No breed or sex predilections for CAE were noted in a variety of studies of CAEV-infected goats from several countries (CRAWFORD and ADAMS 1981; ELLIS et al. 1983a; ZWAHLEN et al. 1983; ADAMS et al. 1984; CAPORALE et al. 1985; DAWSON and WILESMITH 1985; EAST et al. 1987). However, in one study where purebred Saanen goat kids were orally infected with the same dose of biologically cloned isolates of CAEV and observed over a 3-year period, differences in the incidence and frequency of clinical arthritis and the incidence and severity of lesions were noted within groups given the same virus (CHEEVERS et al. 1988). The reason for the difference in disease expression among different goats is not known.

5 Effect of CAEV Genotype on Disease

5.1 Virulence

Genetically distinct CAEV isolates have different virulence. This was demonstrated in goats infected by the oral route with two biologically cloned isolates of CAEV which caused a different incidence and severity of disease (CHEEVERS et al. 1988). The CAEV 63 and CAEV Co isolates exhibited genetic divergence as determined by restriction endonuclease maps (PYPER et al. 1984; ROBERSON and CHEEVERS 1984). Even though the isolates shared antigenic determinants on all structural proteins (GOGOLEWSKI et al. 1985), they could be differentiated by monoclonal antibodies to p28 epitopes (MCGUIRE et al. 1987) and by neutralizing antibody (MCGUIRE et al. 1988) which presumably recognized gp135. Three years after oral inoculation, 17 CAEV 63-infected goats had arthritis in 34 of 34 radiocarpal joints while 18 CAEV Co-infected goats had arthritis in 25 of 36 radiocarpal joints (CHEEVERS et al. 1988). This difference also occurred in other joints; 31 of 34 CAEV 63 and 18 of 36 CAEV Co-infected femorotibial joints had arthritis. Also, the arthritis was clinically and pathologically more severe in the goats with CAEV 63 infection. Similar observations were made with phenotypically distinct ovine lentivirus strains (LAIRMORE et al. 1987) which had different in vivo pathogenicity (LAIRMORE et al. 1988). The reason for the effect of lentivirus genotype on virulence is not clear, although several possibilities are emerging from studies on various lentiviruses (SODROSKI et al. 1985; FISHER et al. 1986).

5.2 Antigenic Variation of Neutralization-Sensitive Epitopes

Antigenic variation of lentiviruses was first described with EIAV (KONO et al. 1973) and later with visna virus (NARAYAN et al. 1977). Antigenic variation of these lentiviruses occurred in the inoculated hosts, and its demonstration required inoculation with homogeneous virus, demonstration of serum-neutralizing antibody to inoculated virus, and subsequent isolation of a virus variant which was not neutralized by the serum antibody existing before or at the time of the virus variant isolation. Evaluation of in vivo antigenic variation of CAEV was hampered by the previous failure to demonstrate serum-neutralizing antibodies to CAEV in infected goat serum samples (KLEVJER-ANDERSON and McGUIRE 1982; NARAYAN et al. 1984). Simultaneous intravenous injection of large quantities of CAEV and inactivated *Mycobacterium tuberculosis* into two CAEV-infected goats induced serum-neutralizing antibodies (NARAYAN et al. 1984), indicating that the goat had the lymphocyte repertoire to make CAEV-neutralizing antibody. Desialylation of CAEV enhanced the kinetics of virus neutralization by these goat antibodies (HUSO et al. 1988). Subsequently, neutralizing antibody was demonstrated in serum from goats experimentally infected with CAEV in Australia (ELLIS et al. 1987). In this report, neutralizing activity to the inoculating virus was detectable by 18 weeks p.i., and an antigenic variant of the inoculated virus was isolated at 78 weeks p.i. from peripheral blood leukocytes by cocultivation on susceptible cells (ELLIS et al. 1987). In another study, serum-neutralizing antibodies to CAEV 63 were demonstrated in CAEV 63-infected goats (McGUIRE et al. 1988). Demonstration of serum-neutralizating antibodies in the latter study, in contrast to earlier failures (KLEVJER-ANDERSON and McGUIRE 1982), was attributed to evaluating sera later in infection and the use of a more sensitive neutralizing assay. Appearance of neutralizing antibodies occurred months or years after infection and the neutralization titers were relatively low (McGUIRE et al. 1988). Antigenic variants of CAEV 63 were isolated from cell-free carpal synovial fluid aspirated from four goats at 9–18 months p.i. (McGUIRE et al. 1988). These joints were inflamed at the time the CAEV antigenic variants were isolated. Two of these virus variants were evaluated in an earlier study and were not distinguished by a neutralizing rabbit serum which was complement dependent (ANDERSON et al. 1983b).

5.3 Antigenic Variants and Disease Progression

Consideration of the possible role of antigenic variation in the progression of CAEV arthritis and other lesions requires an assessment of antigenic variation in other lentiviral diseases. In EIA, the virus antigenic variants were isolated from cell-free serum and were isolated during episodes of clinical disease (KONO et al. 1973; GUTEKUNST and BECVAR 1979; MONTELARO et al. 1984). The EIAV antigenic variants were defined by changes in neutralization-sensitive epitopes which were correlated with nucleic acid sequence changes in the major surface glycoprotein gp90 (PAYNE et al. 1987). In visna virus infection, initial antigenic variants were isolated from persistently infected sheep by cocultivation of peripheral blood cells with susceptible cells in tissue culture (NARAYAN et al. 1977). Subsequent studies showed nucleic acid changes in the major surface glycoprotein, gp135, which induced neutralizing antibody (SCOTT et al. 1979). These and other studies (NARAYAN et al. 1978) suggested that the appearance of visna virus antigenic variants was involved in the pathogenesis of encephalitis by initiating additional inflammation. Other investigators (LUTLEY et al. 1983; THORMAR et al. 1983) evaluated the appearance of visna virus antigenic variants over long periods and discounted the importance of antigenic variants in pathogenesis of CNS lesions. In one study, 12 of 76 isolates made over a 7-year period by cocultivation of blood, brain, and CSF cells with susceptible cells were antigenic variants (LUTLEY et al. 1983). Since the virus used for inoculation was reisolated in most cases, it was suggested that the antigenic variant did not replace the inoculated virus and was not essential for the development of clinically evident CNS lesions (LUTLEY et al. 1983). In another study, antigenic variants were isolated from blood or synovial fluid cells of sheep infected for 1–3 years by cocultivation with susceptible cells (THORMAR et al. 1983). Antigenic variants were reisolated from cells of sheep which had neutralizing antibodies, and it was concluded that the antigenic variants did not seem to have a role in the pathogenesis of disease (THORMAR et al. 1983).

One problem with interpretation of the role of antigenic variants in visna virus disease is that the variants were isolated from cells by cocultivation and may have represented virus that was restricted from expression in vivo (HAASE et al. 1977). If serum-neutralizing antibodies demonstrated in vitro are effective in vivo, then the only virus that should be present in measurable quantities in the serum is a virus variant not recognized by the serum-neutralizing antibody. Such is the case with EIAV (KONO et al. 1973). Appearance of cell-free antigenic variants in the serum or other fluids should result in the infection of susceptible cells and the production of more virus until neutralizing antibody is formed to the antigenic variant. Infected cells with restricted virus expression or infected cells producing virus which is immediately neutralized are likely present and are likely isolated by cocultivation. However, only those cells producing a mutant virus (antigenic variant) not recognized by neutralizing antibodies can result in cell-free viremia. Therefore, when correlation between the isolation of antigenic variants and the exacerbation of clinical signs or lesions is being made, the best correlation should be between the isolation of cell-free antigenic variants and the presence of signs or lesions. Ability to isolate CAEV antigenic variants from cell-free synovial fluid of arthritic goats (McGuire et al. 1988) provides a system to test the role of antigenic variants in the pathogenesis of tissue lesions.

6 Conclusions

Caprine arthritis encephalitis virus infection causes disease in several organ systems, notably the CNS, joints, mammary gland, and lungs. In experimental disease, lesions are most reproducible in the joints and in the mammary gland. This is especially the case when the natural route of oral infection is used. The outcome of disease is influenced by the genotype of the host and by the genotype of the virus. Mechanisms for host and virus genotype control on infection are unknown. However, it is clear that CAEV can readily change its genotype, even during infection of a single host. Disease in CAE is a result of virus persistence with occasional expression of virus likely because of antigenic variation. Reaction of host immune responses with expressed virus causes inflammation, accounting for one major mechanism of disease.

References

Adams DS, Crawford TB, Banks KL, McGuire TC, Perryman LE (1980a) Immune responses of goats persistently infected with caprine arthritis-encephalitis virus. Infect Immun 28: 421–427
Adams DS, Crawford TB, Klevjer-Anderson P (1980b) A pathogenetic study of the early connective tissue lesions of viral caprine arthritis-encephalitis. Am J Pathol 99: 257–278
Adams DS, Klevjer-Anderson P, Carlson JL, McGuire TC, Gorham JR (1983) Transmission and control of caprine arthritis-encephalitis virus. Am J Vet Res 44: 1670–1675
Adams DS, Oliver RE, Ameghino E, DeMartini, JC, Verwoerd DW, Houwers DJ, Waghela S, Gorham JR, Hyllseth B, Dawson M, Trigo FJ, McGuire TC (1984) Serological evidence of caprine arthritis-encephalitis infection in eleven of fourteen countries tested. Vet Rec 115: 493–495
Anderson LW, Klevjer-Anderson P, Liggitt HD (1983a) Susceptibility of blood-derived monocytes and macrophages to caprine arthritis-encephalitis virus. Infect Immun 41: 837–840
Anderson LW, Klevjer-Anderson P, McGuire TC (1983b) Immune rabbit sera do not recognize antigenic variants of caprine arthritis-encephalitis virus isolated during persistent infection. Infect Immun 42: 845–847
Banks KL, Adams DS, McGuire TC, Carlson JL (1983) Experimental infection of sheep by caprine arthritis-encephalitis virus and goats by progressive pneumonia virus. Am J Vet Res 44: 2307–2311
Banks KL, Jacobs CA, Michaels FH, Cheevers WP (1987) Lentivirus infection augments concurrent antigen-induced arthritis. Arthritis Rheum 30: 1046–1053
Caporale VP, Balbo S, Lelli R, Semproni G, Caccia A, Baldelli R (1985) Investigations on lentivirus infections in Italian caprine population. Zbl Veterinarmed [B] 32: 652–659
Cheevers WP, McGuire TC (1988) The lentiviruses: maedi/visna, caprine arthritis-encephalitis, and equine infectious anemia. Adv Virus Res 34: 189–215

Cheevers WP, Knowles DP, McGuire TC, Cunningham DR, Adams DS, Gorham JR (1988) Chronic disease in goats orally infected with two isolates of the caprine arthritis-encephalitis lentivirus. Lab Invest 58: 510–517
Cheevers WP, Roberson S, Klevjer-Anderson P, Crawford TB (1981) Characterization of caprine arthritis-encephalitis virus: a retrovirus of goats. Arch Virol 67: 111–117
Chiu IM, Yaniv A, Dahlberg JE, Gazit A, Skuntz SF, Tronick SR, Aaronson SA (1985) Nucleotide sequence evidence for relationship of AIDS retrovirus to lentiviruses. Nature 317: 366–368
Cork LC, Davis WC (1975) Ultrastructural features of viral leukoencephalomyelitis of goats. Lab Invest 32: 359–365
Cork LC, Narayan O (1980) The pathogenesis of viral leukoencephalomyelitis-arthritis of goats. I. Persistent viral infection with progressive pathologic changes. Lab Invest 42: 596–602
Cork LC, Hadlow WJ, Crawford TB, Gorham JR, Piper RC (1974a) Infectious leukoencephalomyelitis of young goats. J Infect Dis 129: 134–141
Cork LC, Hadlow WJ, Gorham JR, Piper RC, Crawford TB (1974b) Pathology of viral leukoencephalomyelitis of goats. Acta Neuropathol (Berl) 29: 281–292
Crawford TB, Adams DS (1981) Caprine arthritis-encephalitis: clinical features and presence of antibody in selected goat populations. J Am Vet Med Assoc 178: 713–719
Crawford TB, Adams DS, Cheevers WP, Cork LC (1980a) Chronic arthritis in goats caused by a retrovirus. Science 207: 997–999
Crawford TB, Adams DS, Sande RD, Gorham JR, Henson JB (1980b) The connective tissue component of the caprine arthritis-encephalitis syndrome. Am J Pathol 100: 443–454
Cross RF, Smith CK, Moorhead PD (1975) Vertical transmission of progressive penumonia virus of sheep. Am J Vet Res 36: 465–468
Cutlip RC, Lehmkuhl HD, Jackson TA (1981) Intrauterine transmission of ovine progressive penumonia virus. Am J Vet Res 42: 1795–1797
Dawson M, Wilesmith JW (1985) Serological survey of lentivirus (maedi-visna/caprine arthritis-encephalitis) infection in British goat herds. Vet Rec 117: 86–89
DeMartini JC, Banks KL, Greenlee A, Adams DS, McGuire TC (1983) Augmented T lymphocyte responses and abnormal B lymphocyte numbers in goats chronically infected with the retrovirus causing caprine arthritis-encephalitis. Am J Vet Res 44: 2064–2069
East NE, Rowe JD, Madewell BR, Floyd K (1987) Serologic prevalence of caprine arthritis-encephalitis virus in California goat dairies. J Am Vet Med Assoc 190: 182–186
Ellis T, Robinson W, Wilcox G (1983a) Characterisation, experimental infection and serological response to caprine retrovirus. Aust Vet J 60: 321–326
Ellis T, Robinson W, Wilcox G (1983b) Effect of colostrum deprivation of goat kids on the natural transmission of caprine retrovirus infection. Aust Vet J 60: 326–329
Ellis TM, Carman H, Robinson WF, Wilcox GE (1986) The effect of colostrum-derived antibody on neo-natal transmission of caprine arthritis-encephalitis virus infection. Aust Vet J 63: 242–245
Ellis TM, Wilcox GE, Robinson WF (1987) Antigenic variation of caprine arthritis-encephalitis virus during persistent infection of goats. J Gen Virol 68: 3145–3152
Ellis TM, Robinson WF, Wilcox GE (1988) The pathology and aetiology of lung lesions in goats infected with caprine arthritis-encephalitis virus. Aust Vet J 65: 69–73
Fisher AG, Ratner L, Mitsuya H, Marselle LM, Harper ME, Broder S, Gallo RC, Wong-Staal F (1986) Infectious mutants of HTLV-III with changes in the 3' region and markedly reduced cytopathic effects. Science 233: 655–659
Gates NL, Winward LD, Gorham JR, Shen DT (1978) Serologic survey of prevalence of ovine progressive pneumonia in Idaho range sheep. J Am Vet Med Assoc 173: 1575–1577
Gendelman HE, Narayan O, Molineaux S, Clements JE, Ghotbi Z (1985) Slow, persistent replication of lentiviruses: role of tissue macrophages and macrophage precursors in bone marrow. Proc Natl Acad Sci USA 82: 7086–7090
Gogolewski RP, Adams DS, McGuire TC, Banks KL, Cheevers WP (1985) Antigenic cross-reactivity between caprine arthritis-encephalitis, visna and progressive penumonia viruses involves all virion-associated proteins and glycoproteins. J Gen Virol 66: 1233–1240
Gutekunst DE, Becvar CS (1979) Responses in horses infected with equine infectious anemia virus adapted to tissue culture. Am J Vet Res 40: 974–977
Haase A, Stowring L, Narayan O, Griffin D, Price D (1977) Slow persistent infection caused by visna virus: role of host restriction. Science 195: 175–177
Houwers DJ, König CDW, deBoer GF, Schaake J (1983) Maedi-visna control in sheep. I. Artificial rearing of colostrum-deprived lambs. Vet Microbiol 8: 179–185

Huso DL, Narayan O, Hart GW (1988) Sialic acids on the surface of caprine arthritis-encephalitis virus define the biological properties of the virus. J Virol 62: 1974–1980

Johnson GC, Adams DS, McGuire TC (1983a) Pronounced production of polyclonal immunoglobulin G1 in the synovial fluid of goats with caprine arthritis-encephalitis virus infection. Infect Immun 41: 805–815

Johnson GC, Barbet AF, Klevjer-Anderson P, McGuire TC (1983b) Preferential immune response to virion surface glycoproteins by caprine arthritis-encephalitis virus-infected goats. Infect Immun 41: 657–665

Klevjer-Anderson P, Anderson LW (1982) Caprine arthritis-encephalitis virus infection of caprine monocytes. J Gen Virol 58: 195–198

Klevjer-Anderson P, McGuire TC (1982) Neutralizing antibody response of rabbits and goats to caprine arthritis-encephalitis virus. Infect Immun 38: 455–461

Klevjer-Anderson P, Adams DS, Anderson LW, Banks KL, McGuire TC (1984) A sequential study of virus expression in retrovirus-induced arthritis of goats. J Gen Virol 65: 1519–1525

Kono Y, Kobayashi K, Fukunaga Y (1973) Antigenic drift of equine infectious anemia virus in chronically infected horses. Arch Gesamte Virusforsch 41: 1–10

Koyanagi Y, Miles S, Mitsuyasu RT, Merrill JE, Vinters HV, Chen ISY (1987) Dual infection of the central nervous system by AIDS viruses with distinct cellular tropisms. Science 236: 819–822

Lairmore MD, Rosadio RH, DeMartini JC (1986) Ovine lentivirus lymphoid interstitial pneumonia. Rapid induction in neonatal lambs. Am J Pathol 125: 173–181

Lairmore MD, Akita GY, Russell HI, DeMartini JC (1987) Replication and cytopathic effects of ovine lentivirus strains in alveolar macrophages correlate with *in vivo* pathogenicity. J Virol 61: 4038–4042

Lairmore MD, Poulson JM, Adducci TA, DeMartini JC (1988) Lentivirus-induced lymphoproliferative disease. Comparative pathogenicity of phenotypically distinct ovine lentivirus strains. Am J Pathol 130: 80–90

Lutley R, Petursson G, Pálsson PA, Georgsson G, Klein J, Nathanson N (1983) Antigenic drift in visna: virus variation during long-term infection of Icelandic sheep. J Gen Virol 64: 1433–1440

MacDiarmid SC (1984) Scheme to accredit flocks free from caprine arthritis encephalitis infection. NZ Vet J 32: 165–171

McGuire TC, Adams DS, Johnson GC, Klevjer-Anderson P, Barbee DD, Gorham JR (1986) Acute arthritis in caprine arthritis-encephalitis virus challenge exposure of vaccinated or persistently infected goats. Am J Vet Res 47: 537–540

McGuire TC, Brassfield AL, Davis WC, Cheevers WP (1987) Antigenic and structural variation of the p28 core polypeptide of goat and sheep retroviruses. J Gen Virol 68: 2259–2263

McGuire TC, Norton LK, O'Rourke KI, Cheevers WP (1988) Antigenic variation of neutralization-sensitive epitopes of caprine arthritis-encephalitis lentivirus during persistent arthritis. J Virol 62: 3488–3492

Montelaro RC, Parekh B, Orrego A, Issel CJ (1984) Antigenic variation during persistent infection by equine infectious anemia virus, a retrovirus. J Biol Chem 259: 10539–10544

Narayan O, Cork LC (1985) Lentiviral diseases of sheep and goats: chronic pneumonia, leukoencephalomyelitis and arthritis. Rev Infect Dis 7: 89–98

Narayan O, Griffin DE, Chase J (1977) Antigenic shift of visna virus in persistently infected sheep. Science 197: 376–378

Narayan O, Griffin DE, Clements JE (1978) Visna mutation during 'slow infection': temporal development and characterization of mutants of visna virus recovered from sheep. J Gen Virol 41: 343–352

Narayan O, Clements JE, Strandberg JD, Cork LC, Griffin DE (1980) Biological characterization of the virus causing leukoencephalitis and arthritis in goats. J Gen Virol 50: 69–79

Narayan O, Kennedy-Stoskopf S, Sheffer D, Griffin DE, Clements JE (1983) Activation of caprine arthritis-encephalitis virus expression during maturation of monocytes to macrophages. Infect Immun 41: 67–73

Narayan O, Sheffer D, Griffin DE, Clements J, Hess J (1984) Lack of neutralizing antibodies to caprine arthritis-encephalitis lentivirus in persistently infected goats can be overcome by immunization with inactivated *Mycobacterium tuberculosis*. J Virol 49: 349–355

Norman S, Smith MC (1983) Caprine arthritis-encephalitis: review of the neurologic form in 30 cases. J Am Vet Med Assoc 182: 1342–1345

Oliver RE, Gorham JR, Perryman LE, Spencer GR (1981) Ovine progressive pneumonia: experimental intrathoracic, intracerebral, and intra-articular infections. Am J Vet Res 42: 1560–1564

O'Sullivan BM, Eaves FW, Baxendell SA, Rowan KJ (1978) Leucoencephalomyelitis of goat kids. Aust Vet J 54: 479–483

Payne SL, Fang FD, Liu CP, Dhruva BR, Rwambo P, Issel CJ, Montelaro RC (1987) Antigenic variation and lentivirus persistence: variations in envelope gene sequences during EIAV infection resemble changes reported for sequential isolates of HIV. Virology 161: 321–333

Price RW, Brew B, Sidtis J, Rosenblum M, Scheck AC, Cleary P (1988) The brain in AIDS: central nervous system HIV-1 infection and AIDS dementia complex. Science 239: 586–592

Pyper JM, Clements JE, Molineaux SM, Narayan O (1984) Genetic variation among lentiviruses: homology between visna virus and caprine arthritis-encephalitis virus is confined to the 5′ *gag-pol* region and a small portion of the *env* gene. J Virol 51: 713–721

Pyper JM, Clements JE, Gonda MA, Narayan O (1986) Sequence homology between cloned caprine arthritis encephalitis virus and visna virus, two neurotropic lentiviruses. J Virol 58: 665–670

Roberson SM, Cheevers WP (1984) A physical map of caprine arthritis-encephalitis provirus. Virology 134: 489–492

Robinson WF, Ellis TM (1986) Caprine arthritis-encephalitis virus infection: from recognition to eradication. Aust Vet J 63: 237–241

Scott JV, Stowring L, Haase AT, Narayan O, Vigne R (1979) Antigenic variation in visna virus. Cell 18: 321–327

Sims LD, Hale CJ, McCormick BM (1983) Progressive interstitial pneumonia in goats. Aust Vet J 60: 368–371

Smith VW, Dickson J, Coackley W, Maker D (1981) Caprine syncytial retroviruses. Aust Vet J 57: 481–483

Sodroski J, Rosen C, Wong-Staal F, Salahuddin SZ, Popovic M, Arya S, Gallo RC, Haseltine WA (1985) *Trans*-acting transcriptional regulation of human T-cell leukemia virus type III long terminal repeat. Science 227: 171–173

Sonigo P, Alizon M, Staskus K, Klatzmann D, Cole S, Danos O, Retzel E, Toillais P, Haase A, Wain-Hobson S (1985) Nucleotide sequence of the visna lentivirus: relationship to the AIDS virus. Cell 42: 369–382

Sundquist B, Jönsson L, Jacobsson SO, Hammarberg KE (1981) Visna virus meningoencephalomyelitis in goats. Acta Vet Scand 22: 315–330

Surman PG, Daniels E, Dixon BR (1987) Caprine arthritis-encephalitis virus infection of goats in South Australia. Aust Vet J 64: 266–269

Thormar H, Barshatzky MR, Arnesen K, Kozlowski PB (1983) The emergence of antigenic variants is a rare event in long-term visna virus infection *in vivo*. J Gen Virol 64: 1427–1432

Woodard JC, Gaskin JM, Poulos PW, MacKay RJ, Burridge MJ (1982) Caprine arthritis-encephalitis: clinicopathologic study. Am J Vet Res 43: 2085–2096

Zwahlen R, Aeschbacher T, Baker M, Stucki M, Wyder-Walter M, Weiss M, Steck F (1983) Lentivirus Infektionen bei Ziegen mit Carpitis und interstitieller Mastitis. Schweiz Arch Tierheilkd 125: 281–299

Retroviral Infections of the CNS of Nonhuman Primates

A. A. LACKNER, L. J. LOWENSTINE, and P. A. MARX

1 Introduction

Central nervous system infection by retroviruses has become a topic of great concern due to the association of neurologic sequelae with infection by the human immunodeficiency virus (HIV) (CARNE et al. 1985; HO et al. 1985; LEVY et al. 1985; PETITO et al. 1985; PRICE et al. 1988; RESNICK et al. 1985; SHAW et al. 1985; VAZEUX et al. 1988) and the human T-cell leukemia virus type 1 (HTLV-1) (JACOBSON et al. 1988; PICCARDO et al. 1988; POIESZ et al. 1980). The epidemiology, pathology, and virology of these neurologic sequelae are the subject of other chapters in this book. Although encephalopathy and dementia occur frequently in association with HIV infection of the brain in children and adults (HO et al. 1987; PRICE et al. 1988), many questions remain unanswered about the pathogenesis of this CNS infection. Similarly, HTLV-1 has been strongly associated with a slowly progressive myelopathy known as tropical spastic paraparesis (JACOBSON et al. 1988; BHAGAVATI et al. 1988) but a causal relationship has not been firmly established. In addition, recent but as yet unconfirmed reports implicate a retrovirus related to HTLV-1 in multiple sclerosis (KARPAS et al. 1986; KOPROWSKI et al. 1985; KOPROWSKI and DeFreitas 1988; OHTA et al. 1986).

California Primate Research Center and Department of Veterinary Pathology, University of California, Davis, CA 95616, USA

Relevant animal models particularly in nonhuman primates would help in understanding the pathogenesis of retroviral CNS involvement. Lentivirus diseases of sheep and goats and a spongiform polioencephalomyelopathy in wild mice discussed elsewhere in this volume have set the precedents in nature for nononcogenic retrovirus infections of the CNS. However, none of these are optimal animal models of HIV- or HTLV-1-associated neurologic disease. Retroviral infections of the CNS of nonhuman primates may provide better animal models due to the close evolutionary relationship between nonhuman primates and man and their respective retroviruses. We will briefly review the known retroviruses of nonhuman primates and then concentrate on those retroviruses which have been identified in the CNS with special emphasis on the nononcogenic, immunosuppressive retroviruses.

2 Overview of Retroviruses in Nonhuman Primates

After the discovery of murine leukemia virus in the early 1950s (GROSS 1951) frequent attempts were made to isolate and identify retroviruses in human and nonhuman primates. Although retroviruses were detected and often isolated from many vertebrates, the first primate retrovirus was not reported until 1970. This and all subsequent isolates from nonhuman primates belong to one of four retroviral subfamilies: type C oncovirus, type D retrovirus, spumavirus, and lentivirus (Table 1). The first primate retroviral isolate, designated Mason-Pfizer monkey virus (MPMV), was detected in and isolated from a mammary carcinoma of an 8-year-old female rhesus monkey (*Macaca mulatta*) (CHOPRA and MASON 1970; JENSEN et al. 1970). At about the same time, type C oncoviruses were identified in a fibrosarcoma of a pet woolly monkey (*Lagothrix* sp.) (THEILEN et al. 1971) and isolated from spontaneous lymphosarcomas of gibbons (*Hylobates lar*) (KAWAKAMI et al. 1972; SNYDER et al. 1973). These retroviral isolates (simian sarcoma virus and gibbon ape leukemia virus) were found to be closely related (DEINHARDT 1980). In the following years a variety of type C and D endogenous retroviruses were isolated from normal baboons (BaEV) (TODARO et al. 1976), stump-tailed macaques (MAC-1) (TODARO et al. 1978b), rhesus macaques (MMC-1) (RABIN et al. 1979), a colobus monkey (CPC-1) (SHERWIN and TODARO 1979), tree shrews (TRV-1) (FLÜGEL et al. 1978), owl monkeys (OMC-1) (TODARO et al. 1978c), spectacled langurs (PO-1-Lu) (TODARO et al. 1978a), and squirrel monkeys (SMRV) (HEBERLING et al. 1977).

After the discovery of human T-cell leukemia virus type 1 (HTLV-1) in 1980, the first human retrovirus, the search for additional primate retroviruses intensified. A new type C retrovirus closely related to HTLV-1 was isolated from several Asian and African nonhuman primate species including: chimpanzee (*Pan troglodytes*), African green monkey (*Cercopithecus aethiops*), Sykes' monkeys

Table 1. Retroviruses of nonhuman primates

	Isolate	Species of isolation	References
Type C oncoviruses			
Endogenous	BaEV	Various *Papio* species, baboons	Todaro et al. (1976)
	MAC-1	*Macaca arctoides*, stump-tailed macaque	Todaro et al. (1978b)
	MMC-1	*Macaca mulatta*, rhesus macaque	Rabin et al. (1979)
	OMC-1	*Aotus trivirgatus*, owl monkey	Todaro et al. (1978c)
	CPC-1	*Colobus polykomos*, colobus monkey	Sherwin and Todaro (1979)
	TRV-1	*Tupaia belangeri*, tree shrew	Flügel et al. (1978)
Exogenous	Simian sarcoma virus/ simian sarcoma-associated virus (SSV/SSAV)	*Lagothrix* sp. woolly monkey	Theilen et al. (1971)
	Gibbon ape leukemia virus (GaLV)	*Hylobates lar*, white-handed gibbon	Kawakami et al. (1972)
	Simian T-cell leukemia virus type I (STLV-1)	*Pan troglodytes*, chimpanzee;	
		Cercopithecus aethiops, African green monkey;	Komuro et al. (1984)
		Cercopithecus mitis albogularis, Sykes' monkey;	Ishikawa et al. (1987)
		Hylobates syndactylus, siamang;	Yamamoto et al. (1983)
		and several *Macaca* species	
Spumaviruses	Simian foamy virus (ten serotypes)	Multiple species of nonhuman primates from prosimians through great apes	Hooks and Gibbs (1975)
			Rhodes-Feuillette et al. (1987)
Type D retroviruses	Squirrel monkey retrovirus (SMRV)	*Saimiri sciureus*, squirrel monkey	Herberling et al. (1977)
Endogenous	PO-1-LU	*Presbytis obscuris*, spectacled langur monkey	Todaro et al. (1978a)
Exogenous	Mason-Pfizer monkey virus (MPMV)	*Macaca mulatta*, rhesus monkey	Chopra and Mason (1970)
	Simian AIDS retrovirus serotypes 1 and 2	Multiple *Macaca* species	Marx et al. (1984, 1985)
Lentiviruses	Simian immunodeficiency viruses (SIV) from multiple species		
	SIVmac	*Macaca mulatta*, rhesus macaque	Daniel et al. (1985)
		Macaca fascicularis, cynomolgus macaque	Daniel et al. (1988a)
	SIVagm	*Cercopithecus aethiops*, African green monkeys	Ohta et al. (1988)
			Daniel et al. (1988b)
	SIVsm	*Cercocebus atys*, sooty mangabeys	Fultz et al. (1986)
			Lowenstine et al. (1986)
			Murphey-Corb et al. (1986)
	SIVstm	*Macaca arctoides*, stump-tailed macaque	Lowenstine et al. (1987a, b)
	SIVmne	*Macaca nemestrina*, pig-tailed macaque	Benveniste et al. (1986)
	SIVmnd	*Papio sphinx*, mandrill	Tsujimoto et al. (1988)

(*Cercopithecus mitis albogularis*), *siamang* (*Hylobates syndactylus*[1]), and several species of macaques (*Macaca* sp.) (KOMURO et al. 1984; ISHIKAWA et al. 1987; YAMAMOTO et al. 1983). This virus has been designated simian T-cell leukemia virus type 1 (STLV-1) due to its close relationship to HTLV-1 (KOMURO et al. 1984; WATANABE et al. 1985). The animals from which these isolates were obtained were clinically normal except for one African green monkey with lymphoma resembling adult T-cell lymphoma of humans (TSUJIMOTO et al. 1985; SAKAKIBARA et al. 1986). An additional T-cell lymphoma has been described in a captive lowland gorilla (*Gorilla gorilla*) that was seropositive for STLV-1 (LEE et al. 1985). Southern blot analysis of the lymphoma revealed STLV-1 proviral sequences within the tumor cells (SRIVASTAVA et al. 1986). Three other gorillas seropositive for STLV-1 developed a syndrome of wasting and apparent immunosuppression (BLAKESLEE et al. 1987). Whether or not a causal relationship exists between STLV-1 infection and this wasting syndrome has not been established. Other clinically normal adult African green monkeys naturally infected with STLV-1 have had increased absolute lymphocyte counts and increased percentages of atypical lymphocytes (NODA et al. 1986). Given the molecular and biological similarities between STLV-1 and HTLV-1 and the emergence of tropical spastic paraparesis associated with HTLV-1, further studies are needed to determine if STLV-1 is present in the CNS of infected monkeys and whether they will develop neurologic disease that could be a model of tropical spastic paraparesis.

More recently the recognition of an acquired immune deficiency syndrome in captive macaques (simian AIDS, SAIDS) (HENRICKSON et al. 1983; LETVIN et al. 1983) with many similarities to human AIDS has resulted in the isolation and identification of several new nononcogenic, immunosuppressive retroviruses from these monkeys (see below). These include several new serotypes of type D retrovirus related to MPMV and lentiviruses related to the HIV known as the simian immunodeficiency virus (SIV, formerly STLV III). Except for the retroviruses associated with SAIDS relatively little work has been done to determine whether or not these retroviruses are present in the CNS.

3 Retroviruses in the CNS of Nonhuman Primates

Excluding the endogenous retroviruses belonging to the type C and D subfamilies which are present as multiple copies of cellular DNA per cell in all tissues of the body including the CNS, only five exogenous retroviruses have been detected in the CNS (Table 2). These include: SSV-1/SSAV-1 and GaLV, both

[1] The common and scientific names used in this paper are those used by the authors of the original paper and do not always agree with current nomenclature

Table 2. Retroviral infections of the CNS of nonhuman primates

Virus	Main features
Simian sarcoma virus (SSV/SSAV)	No CNS lesions in the woolly monkey (*Lagothrix* sp.) from which the virus was originally isolated. Intracerebral inoculation across species lines into newborn *Saguinus nigrocollis* resulted in gliomas
Gibbon ape leukemia virus (GaLV)	No CNS lesions. Several isolates have been obtained from brain tissue of normal animals. Source of CNS virus is not known
Spumaviruses	Multiple isolates from the CNS of chimpanzees. No CNS lesions
Type D SAIDS retroviruses (SRV-1)	Latent CNS infection without any recognized pathology. Viral DNA and RNA are present in unidentified brain parenchymal cells without detectable antigen. Virus is shed into the CSF from infected choroid plexus epithelial cells. No alterations in the CSF occur and an intrathecal immune response is not detectable
Simian immunodeficiency virus (SIV)	Viral nucleic acid, antigen, and particles are present in perivascular macrophage and multinucleate giant cell infiltrates in the CNS. Neuropathologic features are very similar to those of AIDS encephalopathy

type C oncoviruses, several serotypes of type D retrovirus, several serotypes of spumavirus, and the lentivirus known as SIV.

3.1 Type C Oncoviruses in the CNS

3.1.1 Simian Sarcoma Virus

Simian sarcoma virus was a one time only isolate from a fibrosarcoma of a pet woolly monkey (Deinhardt 1980; Deinhardt et al. 1973; Kawakami et al. 1973; Wolfe et al. 1972). This sole isolate is similar to other acute transforming sarcoma viruses and was found to consist of two separate virions, a defective transforming virus (simian sarcoma virus, SSV-1), and a replication competent helper virus (simian sarcoma-associated virus, SSAV-1) (Wolfe et al. 1972; Deinhardt 1980). The CNS of the woolly monkey from which SSV-1/SSAV-1 was isolated was not examined for virus and no neurologic lesions were reported. The CNS isolates of SSV-1/SSAV-1 were obtained from newborn marmosets (*Saguinus nigricollis*) subsequent to intracerebral inoculation of SSV-1/SSAV-1 (Johnson et al. 1975; Deinhardt 1980). In this study six of ten newborn marmosets inoculated intracerebrally with 0.1 ml SSV-1/SSAV-1 [2.0×10^3 focus-forming units (FFU)] developed gliomas. Type C virus particles were identified in two brain tumors examined by electron microscopy and virus produced from tumor cell lines was indistinguishable from the stock virus. Virus was also isolated from brain tissue distant from the tumor in one animal and from the CSF in one animal. In three other animals with gliomas, however, virus was not isolated from the CSF. CSF collected at necropsy from three animals with gliomas contained SSV-1/SSAV-1 antibodies, but titers were six- to eightfold lower than in sera.

3.1.2 Gibbon Ape Leukemia Virus

Several isolates of gibbon ape leukemia virus were obtained from spontaneous hematopoietic neoplasms in captive gibbons. The first isolates (GaLV-SF) were from two gibbons with spontaneous lymphosarcomas (KAWAKAMI et al. 1972; SNYDER et al. 1973) that were used for radiographic studies on aging of the spinal column at the San Francisco Medical Center. Similar isolates named GaLV-SEATO were obtained from two gibbons with chronic granulocytic leukemia at the SEATO Medical Research Laboratory in Bangkok, Thailand (DEPAOLI et al. 1973; KAWAKAMI and BUCKLEY 1974). These reports do not describe any neurologic abnormalities or attempted viral isolations from the CNS. An additional isolate (GaLV-H) was derived from a gibbon with lymphocytic leukemia that was part of a colony on Hall's island off Bermuda (GALLO et al. 1978). No CNS abnormalities were observed in this animal but GaLV was isolated from the brain. Low levels of viral RNA were also detected in the brain but viral DNA was not found. It was uncertain if this isolate of GaLV was from the brain parenchyma or from the plasma, which contained an estimated 6×10^7 virions/ml as judged by electron microscopy of negatively stained samples. Three additional isolates of GaLV (GaLV-Br) were obtained by cocultivation of the brains of three healthy gibbons (TODARO et al. 1975). Two of these animals had been previously inoculated with brain extracts from human kuru patients and the third was from an uninoculated cagemate. Cocultivation of the brains of two other gibbons did not result in any detectable retrovirus. At necropsy the brains of all five animals were anatomically, histologically, and ultrastructurally normal. While no evidence of neurologic disease has been reported with GaLV infection, the possibility of chronic degenerative CNS disease occurring after years of infection has not been addressed. Further research to determine the neurologic status of long-term GaLV-infected gibbons and the identity of the cells harboring the virus in the CNS are warranted.

3.2 Spumaviruses

Multiple serotypes of spumavirus (foamy viruses) have been isolated from a wide variety of tissues in several animal species including nonhuman primates and humans (HOOKS and DETRICK-HOOKS 1981; HOOKS and GIBBS 1975; WEISS et al. 1982). Of particular interest is their regular presence in the brains of chimpanzees (ROGERS et al. 1976). Although not associated with any disease, a transient depression in cell-mediated immunity has been demonstrated in rabbits experimentally inoculated with simian foamy virus serotype 7 isolated from a chimpanzee (HOOKS and DETRICK-HOOKS 1979) and human cells have been transformed by simian spumavirus serotype 10 (RHODES-FEUILLETT et al. 1987). It is important to be aware of these agents so as not to confuse them with other more pathogenic retroviruses. Indeed, simian foamy virus has been isolated together with HIV-1 from lymphocytes of HIV-1-infected chimpanzees and the foamy virus has identical reverse transcriptase activity and causes similar cytopathic effects in H9 cells (NARA et al. 1987). Further investigations may yet link the spumaviruses to disease in animals and man.

3.3 Nononcogenic Immunosuppressive Retroviruses in the CNS

At the onset of the AIDS epidemic it was apparent that macaque monkeys at several National Institutes of Health primate research centers (HENRICKSON et al. 1983; LETVIN et al. 1983) had a naturally occurring acquired immune deficiency syndrome remarkably similar to the human disease. Because the spontaneous disease was a fatal immunosuppressive disease with clinical and pathologic features similar to AIDS, this disease was named simian AIDS or "SAIDS" (LONDON et al. 1983). Retroviruses from two subfamilies, type D retroviruses and lentiviruses, have been causally associated with this syndrome (BENVENISTE et al. 1986; DANIEL et al. 1984, 1985; LOWENSTINE et al. 1986; MARX et al. 1984, 1985; MURPHEY-CORB et al. 1986; STROMBERG et al. 1984). The disease has been shown to be readily reproducible by experimental inoculation with body fluids as well as purified virus (GRAVELL et al. 1984; LERCHE et al. 1986; LETVIN et al. 1985; MARX et al. 1984; MAUL et al. 1986; MURPHEY-CORB et al. 1986) and even molecularly cloned virus in the case of the type D retrovirus (HEIDECKER et al. 1987). Both of these immunosuppressive retroviruses have been detected in the CNS but only SIV has been consistently associated with neurologic lesions. The major features of these retroviruses are summarized in Table 2 and discussed in detail below.

3.3.1 Type D Retroviruses

Type D retroviruses are widespread in captive macaques accounting for about 99% of the cases of SAIDS (DANIEL et al. 1988a; LERCHE et al. 1986, 1987; MAUL et al. 1986; GARDNER et al. 1988). Although not closely related to the retroviruses that cause AIDS in humans (BRYANT et al. 1985) and less restricted in cell tropism (MAUL et al. 1988), infection leads to depletion of both B- and T-lymphocytes (MAUL et al. 1985, 1988) and a disease closely resembling that of AIDS (OSBORN et al. 1984). The type D subfamily of retroviruses currently consists of three serotypes (SRV-1, SRV-2, and MPMV) which have been molecularly cloned and totally sequenced (POWER et al. 1986; SONIGO et al. 1986; THAYER et al. 1987). Two additional serotypes have recently been discovered (P. MARX, unpublished). Serotype 1 includes two isolates designated SAIDS retrovirus serotype 1 (SRV-1) from the California Primate Research Center and D-398 from the New England Primate Research Center. Serotype 1 viruses are associated with fatal SAIDS in several species of macaques at these two primate research centers (DANIEL et al. 1984; MARX et al. 1984). Serotype 2 includes SRV-2 and SAID/D/Washington, which cause fatal SAIDS associated with retroperitoneal fibromatosis in four species of macaques at the Oregon and Washington Primate Research Centers (MARX et al. 1985; STROMBERG et al. 1984). Serotype 3 contains the prototype type D retrovirus, Mason-Pfizer monkey virus (Bryant et al. 1986a; Chopra and Mason 1970). This retrovirus has not been associated with current outbreaks of SAIDS but has been shown to be capable of inducing simian AIDS (Bryant et al. 1986a; Fine et al. 1975).

Proof that type D retroviruses are the predominant etiologic agent of SAIDS in macaques comes from seroepidemiologic data (Daniel et al. 1988a; Lerche et al. 1986), induction of SAIDS with molecularly cloned SRV-1 (Heidecker et al. 1987), and protection against SRV-1 challenge with a formalin-inactivated SRV-1 vaccine (Marx et al. 1986). SAIDS induced by type D retroviruses is characterized by generalized lymphadenopathy, splenomegaly, hematologic abnormalities, and a wide variety of opportunistic infections including cytomegalovirus as the most common opportunist (MacKenzie et al. 1986; Meyer et al. 1985; Osborn et al. 1984). The mortality rate in juvenile rhesus monkeys from SRV-1 infection is about 50% (Heidecker et al. 1987; Lerche et al. 1987; Maul et al. 1986).

In SRV-1-infected macaques the virus is widespread with an affinity for germinal centers of lymphoid organs, secretory epithelial cells, and epithelial cells in the germinative zones of the digestive tract (Lackner et al. 1988 and unpublished data). SRV-1 also has a broad cell tropism for rhesus B cells, T-helper (CD4+) and suppressor (CD8+) cells, macrophages and fibroblasts, as well as human B- and T-cell lines (Maul et al. 1988). The cellular receptor for type D retroviruses is unknown but based on their broad cell tropism must be present on many cell types.

Neurologic signs and lesions directly attributable to SRV-1 infection of the CNS have not been recognized (Maul et al. 1986; Osborn et al. 1984). However, occasional animals with SAIDS at the California Primate Research Center have had meningitis secondary to septicemia and one animal housed outdoors had a polioencephalomyelitis thought to be due to infection with western equine encephalitis virus. These CNS lesions have occurred in less than 1% of the animals dying with type-D-associated SAIDS. While primary retroviral CNS disease has not been associated with type D SAIDS, SRV-1 RNA and/or DNA have been demonstrated in the cerebral cortex of four of six rhesus monkeys infected with SRV-1 in the absence of detectable viral antigen, viral particles, or inflammatory cells (Lackner et al. 1986, 1988). In these same animals, viral DNA, proteins, particles and infectious virus could be demonstrated in lymphoid organs, salivary glands, and saliva. The neural cell types harboring the viral nucleic acid are currently unidentified. The lack of inflammation in the brain suggested that the SRV-1 genome-positive cells in the brain were resident cells and not inflammatory cells that had migrated into the CNS from the peripheral circulation. Multiple attempts at viral isolation from the brain parenchyma were unsuccessful. For viral isolation, cerebrum, cerebellum, spleen, and skeletal muscle were collected at necropsy after the animals had been thoroughly exsanguinated. Spleen served as a positive control and skeletal muscle as a negative control. Skeletal muscle has proven to be an excellent internal control in our laboratory, free of infectious virus or viral DNA even when adjacent tissues are positive for infectious virus and viral DNA (BRYANT et al. 1986b). Homogenates of these tissues were cocultivated with Raji cells. In all cases the spleen was positive for virus while cerebrum, cerebellum, and skeletal muscle were negative. The presence of SRV-1 RNA and DNA in the brain in the absence of detectable viral antigen, viral particles, or infectious virus suggested that the virus was latent in the CNS.

Subsequently SRV-1 was isolated from the CSF in 13 of 19 rhesus monkeys without histologic or clinical evidence of neurologic dysfunction (LACKNER et al. 1989). Centrifugation of the CSF prior to culture of the supernatant did not prevent virus isolation. In addition, less than 1 cell/μl of CSF was detected using a hemocytometer and no more than 10 cells/100 μl of CSF were detected on cytospin preparations. These data indicate that SRV-1 isolated from CSF was probably not due to the presence of leukocytes. Infectious virus in the CNS of these animals was limited to the CSF and was not detected in the brain parenchyma by electron microscopy, viral culture, or immunohistochemistry. The apparent source of the virus as shown by immunohistochemistry (Fig. 1) was the choroid plexus, where approximately 1 in 1000 surface epithelial cells contained viral antigen.

Serum and CSF collected at the same time as samples for viral isolation were also assayed for anti-SRV-1 antibodies, albumin, and IgG concentrations and ratios. No altertions in these parameters or antibodies against SRV-1 were detected in the CSF regardless of the presence of SRV-1. This indicated that there was not an increase in permeability of the blood-brain barrier of these animals infected with SRV-1. The absence of neurologic signs and lesions together with the lack of detectable antibodies or changes in the CSF suggested that SRV-1 was innocuous in the CNS of these animals. The absence of neurologic lesions may be

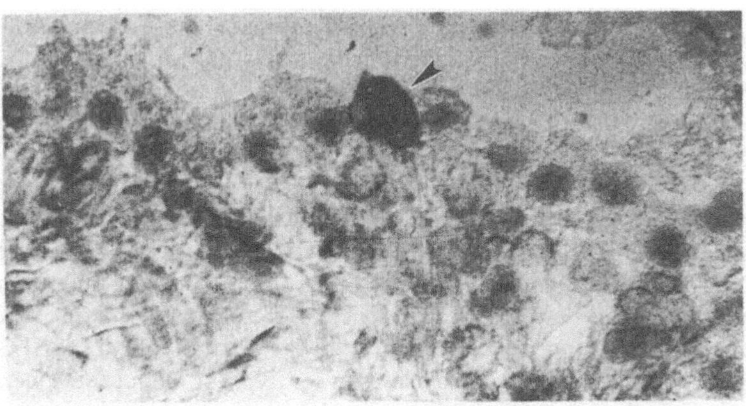

Fig. 1. Localization of SRV-1 in the choroid plexus by immunohistochemistry in rhesus monkeys infected with SRV-1. Approximately 1 in 1000 choroid plexus epithelial cells (*arrowhead*) were found to be positive, × 400

related to the lack of a detectable immune inflammatory response in the CNS. Previously for visna, it has been shown that an immune response to the causative lentivirus was primarily responsible for the pathogenesis of neurologic lesions (NATHANSON et al. 1976). Prevention of the immune response to visna with cyclophosphamide and antithymocyte serum prevented neurologic lesions from occurring but did not effect the frequency of virus isolation from the CNS.

The absence of neurologic signs and lesions in SRV-1-induced SAIDS may also be due, in part, to the level of SRV-1 expression and apparent restriction of productive infection in the CNS to choroid plexus epithelial cells (LACKNER et al. 1988, 1989). From these data we hypothesize that the pathogenesis of SRV-1 infection of the CNS may be as follows: (a) infection with SRV-1 and viremia; (b) productive infection of choroid plexus epithelial cells; (c) shedding of SRV-1 into the CSF by choroid plexus epithelial cells without an intrathecal immune response and; (d) latent infection of the brain parenchyma at a later date. In its apparent latency in the brain parenchyma, restricted expression, and tropism for choroid plexus epithelial cells, SRV-1 infection of the CNS resembles visna virus in sheep. However, in visna, activation of latent CNS virus eventually occurs and leads to neuropathologic manifestations and neurologc disease. Whether this event will yet occur in aging monkeys infected with type D virus or whether the latent virus infection is truly a dead end remains to be determined.

Two other type D retroviruses, MPMV, the prototype type D retrovirus (CHOPRA and MASON 1970; JENSEN et al. 1970), and SAIDS-D/Washington (an isolate from the Washington Primate Research Center) (STROMBERG et al. 1984), have also been detected in the CNS (FINE et al. 1975; TSAI et al. 1987). Each of these type D retroviruses was isolated from the CSF of one animal. None of these animals were reported to have had neurologic signs or lesions. Unlike the present study, however, MPMV was isolated from the brain parenchyma of eight of nine animals and SAIDS-D/Washington was isolated from the brain of one animal. In neither report, however, were negative tissue culture controls described to eliminate the possibility of contamination of cultured tissues by infected peripheral blood mononuclear cells. Excluding the possibility of contamination of tissues by infected peripheral blood mononuclear cells, however, several other possibilities exist to explain the differences in the ability to isolate these type D retroviruses from the brain. The most relevant include the use of different strains of type D retrovirus and the age group of animals infected. SRV-1, MPMV, and SAIDS-D/Washington each belong to a different serotype (BRYANT et al. 1986a, b; MARX et al. 1985) and are genetically distinct (POWER et al. 1986; SONIGO et al. 1986; THAYER et al. 1987). SRV-1 and MPMV are closely related serotypes while SAIDS-D Washington is a more distant relative (serotype 2) showing only about 60% similarity in predicted amino acid sequences of the external envelope gene. This difference in envelope genes may account for the unique ability of SAIDS-D/Washington and other type D retroviruses in serotype 2 to induce retroperitoneal fibromatosis, an aggressive proliferation of fibrous tissue in the abdominal cavity (GIDDENS et al. 1985; STROMBERG et al. 1984). Differences in the envelope could also account for the ability of this virus to infect the brain parenchyma.

Genetic differences could also account for the ability to isolate MPMV from the brain but in addition a different age group of animals was used. In the study of MPMV pathogenesis (FINE et al. 1975), newborn rhesus monkeys were used as compared with juvenile rhesus in other studies. The brain of newborn macaques may be more permissive to viral replication.

3.3.2 Simian Immunodeficiency Virus

While the type D retroviruses are the predominant cause of SAIDS in captive macaques, these viruses are not closely related to HIV and have not been associated with primary CNS disorders. This fact, along with the inability to demonstrate type D retrovirus infection in some cases of SAIDS at the New England Primate Research Center, led to a search for an HIV-related primate lentivirus in rhesus monkeys. This resulted in the isolation and characterization of a virus designated the simian immunodeficiency virus from a rhesus macaque (SIV_{MAC}, formerly STLV-III) (AIDS Subcommittee 1986; DANIEL et al. 1985). A further search in African and Asian primate species by several groups led to the isolation of additional serologically related SIV isolates. These included isolates from African monkeys such as sooty mangabeys (*Cercocebus atys*) and African green monkeys (*Cercopithecus aethiops*) which are apparently healthy carriers (DANIEL et al., 1988b; FULTZ et al. 1986; LOWENSTINE et al. 1986; MURPHEY-CORB et al. 1986; OHTA et al. 1988) and a single isolate from a mandrill (*Papio sphinx*) (TSUJIMOTO et al. 1988). Three other SIV isolates have been obtained from Asian monkeys. Two of these isolates were derived from frozen lymphoid tissue from a stump-tailed macaque (*Macaca arctoides*) at the California Primate Research Center and a pig-tailed macaque (*Macaca nemestrina*) at the Washington Primate Research Center respectively (BENVENISTE et al. 1986; LOWENSTINE et al. 1987a, b). An additional isolate was obtained from a cynomolgus monkey (*Macaca fascicularis*) (DANIEL et al. 1988a). Seroepidemiologic surveys suggest that other species of African monkeys including talapoins (*Miopithecus talapoin*) and De Brazza's monkeys (*Cercopithecus neglectus*) may also be infected with SIV-related retroviruses (LOWENSTINE et al. 1986; OHTA et al. 1988). Serologic and molecular comparisons suggest that all of the isolates of SIV are related to one another and to HIV (BENVENISTE et al. 1986; CHAKRABARTI et al. 1987; FUKASAWA et al. 1988; LOWENSTINE et al. 1986; OHTA et al. 1988; SCHNEIDER and HUNSMANN 1988). The extent of envelope variation, presence or absence of serotypes, and possible differences in pathogenesis remain to be determined.

When inoculated into macaques SIV has induced a fatal AIDS-like disease culminating in profound immune suppression (BENVENISTE et al. 1988; LETVIN et al. 1985; MURPHEY-CORB et al. 1986) and death due to multiple opportunistic infections including disseminated cytomegalovirus infection, adenovirus pancreatitis, oral candidiasis, cryptosporidiosis, trichomoniasis, *Mycobacterium avium-intracellulare* infection, and rarely *Pneumocystis carinii* pneumonia and lymphoma (LETVIN et al. 1985; KING 1986). The immune suppression was characterized by depletion of CD4+ lymphocytes and decreased responses to pokeweed mitogen (a T-cell-dependent B-cell mitogen) (LETVIN et al. 1985). *In vitro*, SIV was shown to be tropic for CD4 + but not CD8 + lymphocytes and that its ability to infect macaque and human lymphocytes could be blocked with anti-CD4 monoclonal antibodies (KANNAGI et al. 1985). Thus, as with HIV the receptor for SIV appears to be the CD4 molecule.

Central nervous system involvement in SIV-induced SAIDS has also been described (BASKIN et al. 1986; BENVENISTE et al. 1988; LETVIN et al. 1985). Opportunistic infections and lesions involving the CNS of macaques with SIV-induced SAIDS have included cytomegalovirus leptomeningitis and ganglioneuritis, toxoplasmosis, lymphoma, and progressive multifocal leukoencephalopathy (Figs. 2 and 3) (GRIBBLE et al. 1975; HUNT et al. 1983; KING 1986; SHARER et al. 1988; STOWELL et al. 1971). A primary retroviral encephalopathy (Fig. 4) similar to what has been described in patients with AIDS has also been reported (BASKIN et al. 1986; BENVENISTE et al. 1988; LETVIN et al. 1985; RINGLER et al. 1988; SHARER et al. 1988). At the New England Primate Research Center 10 of 18 (56%) rhesus monkeys experimentally inoculated with SIV_{MAC} were found at necropsy to have a meningoencephalitis characterized by perivascular infiltrates of macrophages and multinulceate giant cells throughout the brain parenchyma and in the leptomeninges (RINGLER et al. 1988). Spongiosis of adjacent neuropil was frequently seen as well as vascular dilatation and endothelial hypertrophy. Scattered microglial nodules were also present in four of the ten animals. Ultrastructurally, lentiviral particles were present within membrane-bound cytoplasmic vacuoles in macrophages and giant cells. Budding of viral particles

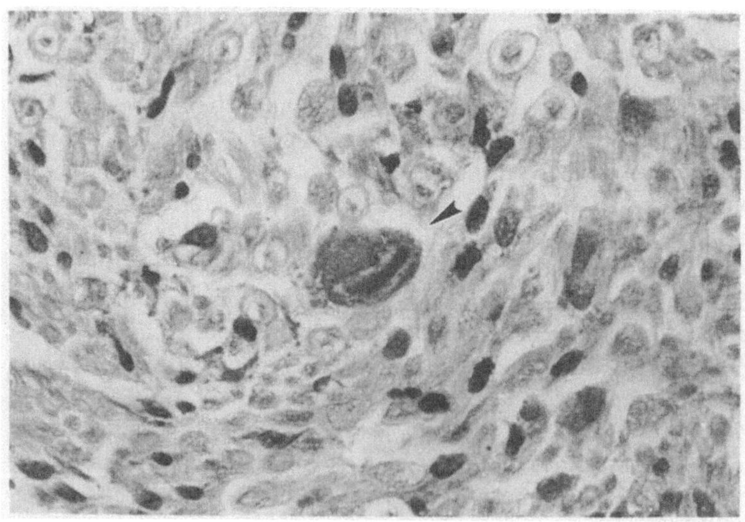

Fig. 2. Neuritis associated with cytomegalovirus infection in a rhesus monkey experimentally infected with SIV. Note inclusion-bearing cell showing cytomegaly (*arrowhead*). H&E, × 350

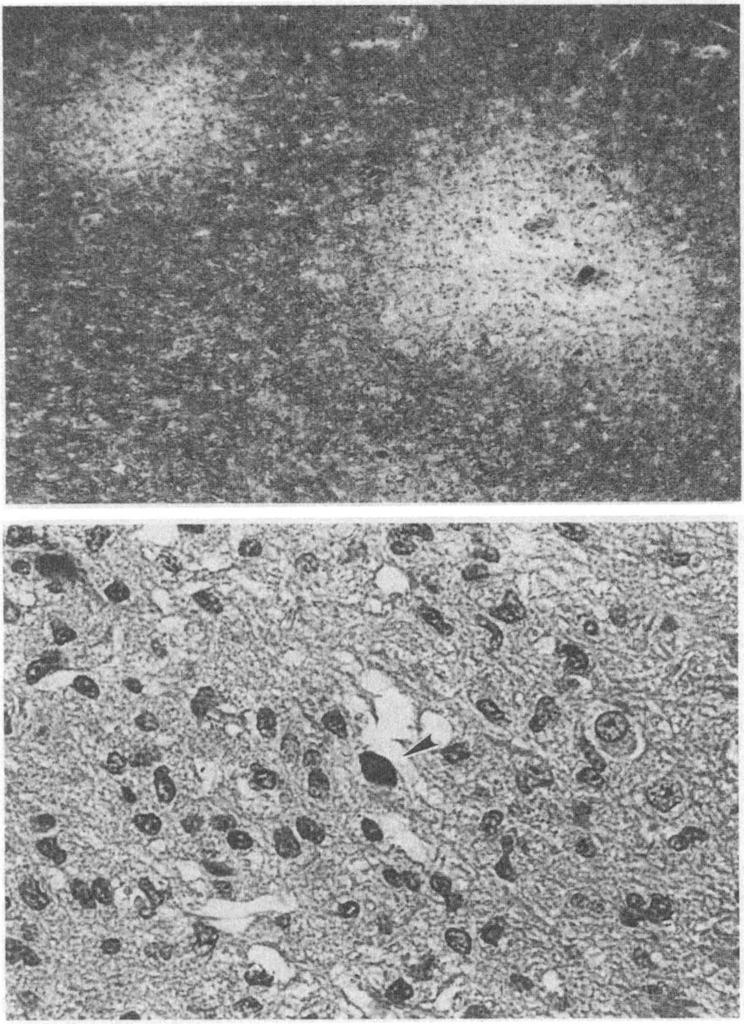

Fig. 3 a, b. Foci of demyelination (**a**) and (**b**) a basophilic nuclear inclusion body (*arrowhead*) typical or progressive multifocal leukoencephalopathy (PML) in an SIV-infected stump-tailed macaque. **a** Luxol fast blue/cresyl violet, × 60; **b** H&E, × 350

into the vacuoles was observed but not from the plasmalemma into the extracellular space.

A very similar encephalitis has been described associated with experimental inoculation of an SIV isolate designated SIV Delta B670 at the Delta Regional Primate Research Center (Sharer et al. 1988). In this study five of five animals that died with SAIDS induced by this isolate had perivascular infiltrates of macrophages and multinucleate giant cells in the leptomeninges and brain

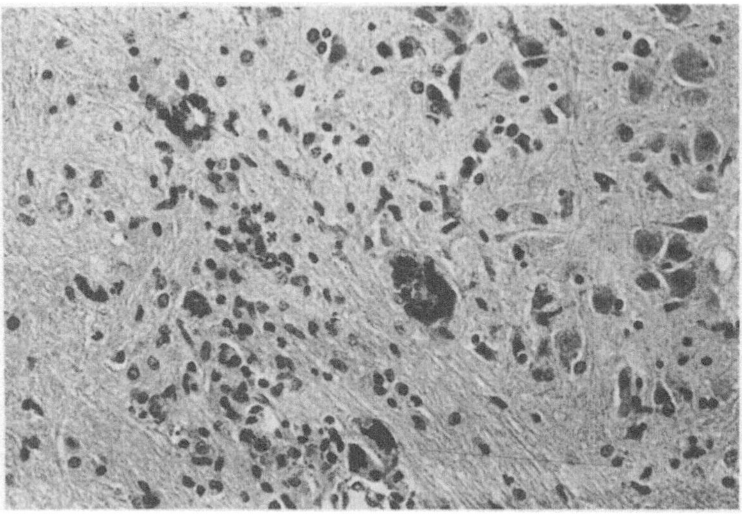

Fig. 4. Photomicrograph of an encephalitis with syncitial giant cells in a stump-tailed macaque infected with SIV. H&E, × 150

parenchyma. Using an SIV-specific nucleic acid probe, polyclonal antisera, and electron microscopy, the SIV genome, antigens, and lentiviral particles were detected in these perivascular infiltrates in the leptomeninges and brain parenchyma. In the white matter a mild diffuse astrocytosis of varying degree was seen in all five animals. Clinically evident neurologic abnormalities were not described in either of these studies. None of the animals with encephalopathy had grossly visible abnormalities of the brain. The encephalitis in these animals was assumed to be a primary retroviral disease due to the presence of viral particles and or virus-specific nucleic acid in the infiltrating macrophages and giant cells and the lack of detectable opportunistic infections in the CNS in all but one animal which had a cytomegalovirus leptomeningitis.

In the middle 1970s a similar giant cell encephalitis was recognized in stump-tailed macaques (*Macaca arctoides*) at the California Primate Research Center as one manifestation of an epizootic of immunosuppression, opportunistic infection, and lymphoma (Holmberg et al. 1985a, b). The giant cell encephalitis was originally thought to be due to *Mycobacterium avium* but no organisms could be demonstrated with acid-fast stains. A retrospective seroepidemiologic study of these animals showed a high seroprevalence (approximately 70%) of antibodies to HIV and SIV (Lowenstine et al. 1987a). Inoculation of a juvenile rhesus macaque with a lymph node extract that had been stored frozen for 9 years resulted in the animal seroconverting and the isolation of an additional SIV isolate (Lowenstine et al. 1987a, b). Serial passage of blood into a second rhesus caused death in 7 months with immunosuppression, cytomegalovirus ganglioneuritis, and focal giant cell encephalitis. Since that time we have seen a similar encephalopathy in rhesus monkeys inoculated with SIV from sooty managabeys and with the macaque isolate from the New England Primate Research Center.

In summary, several retroviruses from four different subfamilies have been found to infect the CNS of nonhuman primates. Only one of these retroviruses, the lentivirus, SIV has been found to induce a primary encephalitis. SIV is an important model for both the immunosuppression and encephalopathy induced by HIV. Future work involving experimental infection of susceptible macaques with SIV will be invaluable for elucidating the pathogenesis of retroviral encephalitis particularly as it relates to AIDS.

Acknowledgments. This work supported by Public Health Service grants RR00169 from the Division of Research Resources, National Institutes of Health, RO1-AI20573, from the National Institute of Allergy and Infectious Diseases, and by the Universitywide Task Force on AIDS, University of California.

A. Lackner is a recipient of a Public Health Service Special Emphasis Career Award in Laboratory Animal Science from the Division of Research Resources, RR00039.

References

AIDS Subcommittee, (1986) International committee for the taxonomy of viruses: human immunodeficiency viruses. Science 232: 697

Baskin GB, Martin LN, Rangan SRS, Gormus BJ, Murphey-Corb M, Wolf RH, Soike KF (1986) Transmissible lymphoma and simian acquired immunodeficiency syndrome in rhesus monkeys. JNCI 77: 127–139

Benveniste RE, Arthur LO, Tsai C-C, Sowder R, Copeland TD, Henderson LE, Oroszlan S (1986) Isolation of a lentivirus from a macaque with lymphoma: comparison with HTLV-III/LAV and other lentiviruses. J Virol 60: 483–490

Benveniste RE, Morton WR, Clark EA, Tsai C-C, Ochs HD, Ward JM, Kuller L, Knott WB, Hill RW, Gale MJ, Thouless ME (1988) Inoculation of baboons and macaques with simain immunodeficiency virus/Mne, a primate lentivirus closely related to human immunodeficiency virus type 2. J Virol 62: 2091–2101

Bhagavati S, Ehrlich G, Kula RW, Kwok S, Sninsky J, Udani V, Poiesz J (1988) Detection of human T-cell lymphoma/leukemia virus type I DNA and antigen in spinal fluid and blood of patients with chronic progressive myelopathy. N Engl J Med 318: 1141–1147

Blakeslee JR, McClure HM, Anderson DC, Bauer RM Huff LY, Olsen RG (1987) Chronic fatal disease in gorillas seropositive for simian T-lymphotropic virus I antibodies. Cancer Lett 37: 1–6

Bryant ML, Yamamoto J, Luciw P, Munn R, Marx P, Higgins J, Pedersen N, Levine A, Gardner MB (1985) Molecular comparison of retroviruses associated with human and simian AIDS. Hematol Oncol 3: 187–197

Bryant ML, Gardner MB, Marx PA, Maul DH, Lerche NW, Osborn KG, Lowenstine LJ, Bogden A, Arthur LO, Hunter E (1986a) Immunodeficiency in rhesus monkeys associated with the original Mason-Pfizer monkey virus. JNCI 77: 957–965

Bryant ML, Marx PA, Shiigi SM, Wilson BJ, McNulty WP, Gardner MB (1986b) Distribution of type D retrovirus sequences in tissues of macaques with simian acquired immune deficiency and retroperitoneal fibromatosis. Virology 150: 149–160

Carne CA, Smith AN, Elkington SG, Preston RE, Tededer RS, Sutherland S, Daly HM,

Craske J (1985) Acute encephalopathy coincident with seroconversion for anti-HTLV-III. Lancet ii: 1206–1207

Chakrabarti L, Guyader M, Alizon M, Daniel MD, Desrosiers RC, Tiollais P, Sonigo P (1987) Sequence of simian immunodeficiency virus from macaque and its relationship to other human and simian retroviruses. Nature 328: 543–547

Chopra HC, Mason MM (1970) A new virus in a spontaneous mammary tumor of a rhesus monkey. Cancer Res 30: 2081–2086

Daniel MD, King NW, Letvin NL, Hunt RD, Sehgal PK, Desrosiers RC (1984) A new type D retrovirus isolated from macaques with an immunodeficiency syndrome. Science 223: 602–605

Daniel MD, Letvin NL, King NW, Kannagi M, Sehgal PK, Hunt RD, Kanki PJ, Essex M, Desrosiers RC (1985) Isolation of T-cell tropic HTLV-III-like retrovirus from macaques. Science 228: 1201–1204

Daniel MD, Letvin NL, Sehgal PK, Schmidt DK, Silva DP, Solomon KR, Hodi SF, Ringler DJ, Hunt RD, King NW, Desrosiers RC (1988a) Prevalence of antibodies to 3 retroviruses in a captive colony of macaque monkeys. Int J Cancer 41: 601–608

Daniel MD, Li Y, Naidu YM, Durda PJ, Schmidt DK, Finger CD, Silva DP, MacKey JJ, Kestler HW, Sehgal PK, King NW, Hayami M, Desrosiers RC (1988b) Simian immunodeficiency viruses from African green monkeys. J Virol 62: 4123–4128

Deinhardt F (1980) Biology of primate retroviruses. In: Klein G (ed) Viral oncology. Raven, New York, pp 357–398

Deinhardt F, Wolfe L, Northrop R, Marczynska B, Ogden J, McDonald R, Falk L, Shramek G, Smith R, Deinhardt J (1973) Simian sarcoma virus: oncogenicity, focus assay, presence of associated virus, and comparison with avian and feline sarcoma virus-induced neoplasia in marmoset monkeys. Bibl Haematol 39: 258–262

DePaoli A, Johnsen DO, Noll WW (1973) Granulocytic leukemia in whitehanded gibbons. J Am Vet Med Assoc 163: 624–628

Fine DL, Landon JL, Pienta RJ, Kubicek MT, Valerio MG, Loeb WF, Chopra HC (1975) Responses of infant rhesus monkeys to inoculation with Mason-Pfizer monkey virus materials. J Natl Cancer Inst 54: 651–658

Flügel RM, Zentgraf H, Munk K, Darai G (1978) Activation of an endogenous retrovirus from *Tupaia* (tree shrew). Nature 271: 543–545

Fukasawa M, Miura T, Hasegawa A, Morikawa S, Tsujimoto H, Miki K, Kitamura T, Hayami M (1988) Sequence of simian immunodeficiency virus from African green monkey, a new member of the HIV/SIV group. Nature 333: 457–461

Fultz PN, McClure H, Anderson DC, Swenson RB, Anand R, Srinivasan A (1986) Isolation of a T-lymphotropic retrovirus from naturally infected sooty mangabey monkeys (*Cercocebus atys*). Proc Natl Acad Sci USA 83: 5286–5290

Gallo RC, Gallagher RE, Wong-Staal F, Aoki T, Markham PD, Schetters H, Ruscetti F, Valerio M, Walling MJ, O'Keeffe RT, Saxinger WC, Smith RG, Gillespie DH, Reitz MS (1978) Isolation and tissue distribution of type C virus and viral components from a gibbon ape (*Hylobates lar*) with lymphocytic leukemia. Virology 84: 359–373

Gardner MB, Luciw P, Lerche N, Marx PE (1988) Non-human primate retrovirus isolates and AIDS. In: Perk K (ed) Immunodeficiency disorders and retroviruses. Academic, New York pp 285–299

Giddens WE, Tsai C-C, Morton WR, Ochs HD, Knitter GH, Blakley GA (1985) Retroperitoneal fibromatosis and acquired immunodeficiency syndrome in macaques. Am J Pathol 119: 253–263

Gravell M, London WT, Lecatsas G, Hamilton RS, Houff SA, Sever JL (1984) Transmission of simian acquired immunodeficiency syndrome (SAIDS) with type D retrovirus isolated from saliva or urine. Proc Soc Exp Biol Med 177: 491–494

Gribble DH, Haden CC, Schwartz LW, Henrickson RV (1975) Spontaneous progressive multifocal leukoencephalopathy (PML) in macaques. Nature 254: 602–604

Gross L (1951) "Spontaneous" leukemia developing in C3H mice following inoculation in infancy with AK-leukemic extracts, or AK-embryos. Proc Soc Exp Biol Med 76: 27–32

Heberling RL, Barker ST, Kalter SS, Smith GC, Helmke RJ (1977) Oncornavirus: isolation from a squirrel monkey (*Saimiri sciureus*) lung culture. Science 195: 289–292

Heidecker G, Lerche NW, Lowenstine LJ, Lackner AA, Osborn KG, Gardner MB, Marx PA (1987) Induction of simian acquired immune deficiency syndrome (SAIDS) with a molecular clone of a type D retrovirus. J Virol 61: 3066–3071

Henrickson RV, Maul DH, Osborn KG, Sever JL, Madden DL, Ellingsworth LR, Anderson JH, Lowenstine LJ, Gardner MB (1983) Epidemic of acquired immunodeficiency in rhesus monkeys. Lancet i: 388–390

Ho DD, Rota TR, Schooley RT, Kaplan JC, Allan JD, Groopman JE, Resnick L, Felsenstein D, Andrews CA, Hirsch MS (1985) Isolation of HTLV-III from cerebrospinal fluid and neural tissues of patients with neurologic syndromes related to the acquired immunodeficiency syndrome. N Engl J Med 313: 1493–1497
Ho DD, Pomerantz RJ, Kaplan JC (1987) Pathogenesis of infection with human immunodeficiency virus. N Engl J Med 317: 278–286
Holmberg CA, Henrickson R, Anderson J, Osburn BI (1985a) Malignant lymphoma in a colony of *Macaca arctoides*. Vet Pathol 22: 42–45
Holmberg CA, Henrickson R, Lenninger R, Anderson J, Hayashi L, Ellingsworth L (1985b) Immunologic abnormalities in a group of *Macaca arctoides* with high mortality due to atypical mycobacterial and other disease processes. Am J Vet Res 46: 1192–1196
Hooks JJ, Detrick-Hooks B (1979) Simian foamy virus-induced immunosuppression in rabbits. J Gen Virol 44: 383–390
Hooks JJ, Detrick-Hooks B (1981) Spumavirinae. Foamy virus group infections: comparative aspects and diagnosis. In Kurstak E, Kurstak C (eds) Comparative diagnosis of viral diseases, vol 4, Academic Press, New York pp 599–618
Hooks JJ, Gibbs CJ (1975) The foamy viruses. Bacteriol Rev 39: 169–185
Hunt RD, Blake BJ, Chalifoux LV, Sehgal PK, King NW, Letvin NL (1983) Transmission of naturally occurring lymphoma in macaque monkeys. Proc Natl Acad Sci USA 80: 5085–5089
Ishikawa K, Fukasawa M, Tsujimoto H, Else JG, Isahakia M, Ubhi NK, Ishida T, Takenaka O, Kawamoto Y, Shotake T, Ohsawa H, Ivanoff B, Cooper RW, Frost E, Grant FC, Spriatna Y, Sutarman, Abe K, Yamamoto K, Hayami M (1987) Serological survey and virus isolation of simian T-cell leukemia/lymphotropic virus type I (STLV-I) in non-human primates in their native countries. Int J Cancer 40: 233–239
Jacobson S, Raine CS, Mingioli ES, McFarlin De (1988) Isolation of an HTLV-1-like retrovirus from patients with tropical spastic paraparesis. Nature 331: 540–543
Jensen EM, Zelljadt I, Chopra HC, Mason MM (1970) Isolation and propagation of a virus from a spontaneous mammary carcinoma of a rhesus monkey. Cancer Res 30: 2388–2393
Johnson L, Wolfe LG, Whisler WW, Norton T, Thakkar B, Deinhardt F (1975) Induction of gliomas in marmosets by simian sarcoma virus, type 1 (SSV-1). Abst Proc Am Assoc Cancer Res 16: 119
Kannagi M, Yetz JM, Letvin NL (1985) In vitro growth characteristics of simian T-lymphotropic virus type III. Proc Natl Acad Sci USA 82: 7053–7057
Karpas A, Kampf U, Siden A, Koch M, Poser S (1986) Lack of evidence for involvement of known human retroviruses in multiple sclerosis. Nature 322: 177–178
Kawakami TG, Buckley PM (1974) Antigenic studies on gibbon type C viruses. Transplant Proc 6: 193–196
Kawakami TG, Huff SD, Buckley PM, Dungworth DL, Snyder SP (1972) C-type virus associated with gibbon lymphosarcoma. Nature [New Biol] 235: 170–171
Kawakami TG, Buckley P, Huff S, McKain D, Fielding H (1973) A comparative study in vitro of a simian virus isolated from spontaneous woolly monkey fibrosarcoma and of a known feline fibrosarcoma virus. Bibl Haematol 39: 236–243
King NW (1986) Simian models of acquired immunodeficiency syndrome (AIDS): a review. Vet Pathol 23: 345–353
Komuro A, Watanabe T, Miyoshi, I, Hayami M, Tsujimoto H, Seiki M, Yoshida M (1984) Detection and characterization of simian retroviruses homologous to human T-cell leukemia virus type I. Virology 138: 373–378
Koprowski H, DeFreitas E (1988) HTLV-1 and chronic nervous diseases: present status and a look into the future. Ann Neurol 73 (Suppl): 166–170
Koprowski H, DeFreitas E, Harper ME, Sandberg-Wollheim M, Sheremata WA, Robert-Guroff M, Saxinger CW, Feinberg MB, Wong-Stall F, Gallo RC (1985) Multiple sclerosis and human T-cell lymphotropic retroviruses. Nature 318: 154–160
Lackner AA, Rodriguez MH, Bush CE, Osborn KG, Kwang H-S, Moore PF, Lowenstine LJ, Marx PA, Hinrichs SH, Gardner MB (1986) In situ localization of simian AIDS retrovirous-1 (SRV-1) antigen and RNA. Lab Invest 54: 33A
Lackner AA, Marx PA, Lerche NW, Gardner MB, Kluge JD, Spinner A, Kwang H-S, Lowenstine LJ (1989) Asymptomatic infection of the central nervous system by the macaque immunosuppressive type D retrovirus, SRV-1. J Gen Virol 70: 1641–1651
Lackner AA, Rodriguez MH, Bush CE, Munn RJ, Kwang H-S, Moore PF, Osborn KG, Marx PA, Gardner MB, Lowenstine LJ (1988) Distribution of a macaque immunosuppressive type D retrovirus in neural, lymphoid and salivary tissues. J Virol 62: 2134–2142

Lee RV, Prowten AW, Satchidanand SK (1985) Non-Hodgkin's lymphoma and HTLV-I antibodies in a gorilla. N Engl J Med 312: 118–119

Lerche NW, Osborn KG, Marx PA, Prahalada S, Maul DH, Lowenstine LJ, Munn RJ, Bryant ML, Henrickson RV, Arthur LO, Gilden RV, Barker CS, Hunter E, Gardner MB (1986) Inapparent carriers of simian AIDS type D retrovirus and disease transmission with saliva. JNCI 77: 489–496

Lerche NW, Marx PA, Osborn KG, Maul DH, Lowenstine LJ, Bleviss ML, Moody P, Henrickson RV, Gardner MB (1987) Natural history of endemic type D retrovirus infection and acquired immune deficiency syndrome in group-housed rhesus monkeys. JNCI 79: 847–854

Letvin NL, Eaton KA. Aldrich WR, Sehgal PK, Blake BJ, Schlossman SF, King NW, Hunt RD (1983) Acquired immunodeficiency syndrome in a colony of macaque monkeys. Proc Natl Acad Sci USA 80: 2718–2722

Letvin NL, Daniel MD, Sehgal PK, Desrosiers RC, Hunt RD, Waldron LM, MacKey JJ, Schmidt DK, Chalifoux LV, King NW (1985) Induction of AIDS-like disease in macaque monkeys with T-cell tropic retrovirus STLV-III. Science 230: 71–73

Levy JA, Shimabukuro J, Hollander H, Mills J, Kaminsky L (1985) Isolation of AIDS-associated retroviruses from cerebrospinal fluid and brain of patients with neurological symptoms. Lancet i: 586–588

London WT, Sever JL, Madden DL, Henrickson RV, Gravell M, Maul DH, Dalakas MC, Osborn KG, Houff SA, Gardner MB (1983) Experimental transmission of simian acquired immunodeficiency syndrome (SAIDS) and Kaposi-like skin lesions. Lancet ii: 869–873

Lowenstine LJ, Pedersen NC, Higgins J, Pallis KC, Uyeda A, Marx PA, Lerche NW, Munn RJ, Gardner MB (1986) Seroepidemiologic survey of captive old-world primates for antibodies to human and simian retroviruses and isolation of a lentivirus from sooty mangabeys (*Cercocebus atys*). Int J Cancer 38: 563–574

Lowenstine LJ, Lerche NW, Jennings M, Yee J, Uyeda A, Gardner MB (1987a) Evidence for lentiviral etiology in an epizootic of lymphoma and immunodeficiency in stump-tailed macaques (*Macaca arctoides*) (Abstract). 3rd International conference on AIDS, Washington, DC p 63

Lowenstine L, Lerche N, Marx P, Gardner M, Pedersen N (1987b) An epizootic of simian AIDS caused by SIV in captive macaques in the 1970's. In: Girard M, deThe G, Vallette L (eds) Retroviruses of human AIDS and related animal viruses Pasteur vaccines, Lyon pp 174–176

MacKenzie M, Lowenstine L, Lalchandani R, Lerche N, Osborn K, Spinner A, Bleviss M, Henrickson R, Gardner M (1986) Hematologic abnormalities in simian acquired immune deficiency syndrome. Lab Anim Sci 36: 14–19

Marx PA, Maul DH, Osborn KG, Lerche NW, Moody P, Lowenstine LJ, Henrickson RV, Arthur LO, Gravell M, London WT, Sever JL, Levy JA, Munn RB, Gardner MB (1984) Simian AIDS: isolation of a type D retrovirus and disease transmission. Science 223: 1083–1086

Marx PA, Bryant ML, Osborn KG, Maul DH, Lerche NW, Lowenstine LJ, Kluge JD, Zaiss CP, Henrickson RV, Shiigi SM, Wilson BJ, Malley A, Olson L, McNulty WP, Arthur LO, Gilden RV, Barker CS, Hunter E, Munn RJ, Heidecker G, Gardner MB (1985) Isolation of a new serotype of simian acquired immune deficiency syndrome type D retrovirus from Celebes black macaques (*Macaca nigra*) with immune deficiency and retroperitoneal fibromatosis. J Virol 56: 571–578

Marx PA, Pederson NC, Lerche NW, Osborn KG, Lowenstine LJ, Lackner AA, Maul DH, Kwang H-S, Kluge JD, Zaiss CP, Sharpe V, Spinner A, Allison AC, Gardner MB (1986) Prevention of simian acquired immune deficiency syndrome with a formalin-inactivated type D retrovirus vaccine. J Virol 60: 431–435

Maul DH, Miller CH, Marx PA, Bleviss ML, Madden DL, Henrickson RV, Gardner MB (1985) Immune defects in simian acquired immunodeficiency syndrome. Vet Immunol Immunopathol 8: 201–214

Maul DH, Lerche NW, Osborn KG, Marx PA, Zaiss C, Spinner A, Kluge JD, MacKenzie MR, Lowenstine LJ, Bryant ML, Blakeslee JR, Henrickson RV, Gardner MB (1986) Pathogenesis of simian AIDS in rhesus macaques inoculated with the SRV-1 strain of type D retrovirus. Am J Vet Res 47: 863–868

Maul DH, Zaiss CP, MacKenzie MR, Shiigi SM, Marx PA, Gardner MB (1988) Simian retrovirus D serogroup 1 has a broad cellular tropism for lymphoid and nonlymphoid cells. J Virol 62: 1768–1773

Meyer PR, Ormerod LD, Osborn KG, Lowenstine LJ, Henrickson RV, Modlin RL, Smith RE, Gardner MB, Taylor CR (1985) An immunopathologic evaluation of lympth nodes from monkey and man with acquired immune deficiency syndrome and related conditions. Hematol Oncol 3: 199–210

Murphey-Corb M, Martin LN, Rangan SRS, Baskin GB, Gormus BJ, Wolf RH, Andes WA, West M, Montelaro RC (1986) Isolation of an HTLV-III-related retrovirus from macaques with simian AIDS and its possible origin in asymptomatic mangabeys. Nature 321: 435–437
Nara PL, Robey WG, Arthur LO, Gonda MA, Asher DM, Yanagihara R, Gibbs CJ, Gajdusek DC, Fischinger PJ (1987) Simultaneous isolation of simian foamy virus and HTLV-III/LAV from chimpanzee lymphocytes following HTLV-III or LAV inoculation. Arch Virol 92: 183–186
Nathanson N, Panitch H, Palsson PA, Petursson G, Georgsson G (1976) Pathogenesis of visna: effect of immunosuppression upon early central nervous system lesions. Lab Invest 35: 444–451
Noda Y, Ishikawa K, Sasagawa A, Honjo S, Mori S, Tsujimoto H, Hayami M (1986) Hematologic abnormalities similar to the preleukemic state of adult T-cell leukemia virus. Jpn J Cancer Res 77: 1227–1234
Ohta M, Ohta K, Mori F, Nishitani H, Saida T (1986) Sera from patients with multiple sclerosis react with HTLV-1 and *gag* proteins but not *env* proteins—Western blotting analysis. J Immunol 137: 3440–3443
Ohta Y, Masuda T, Tsujimoto H, Ishikawa K, Morikawa S, Nakai M, Honjo S, Hayami M (1988) Isolation of simian immunodeficiency virus from African green monkeys and seroepidemiologic survey of the virus in various non-human primates. Int J Cancer 41: 115–122
Osborn KG, Prahalada S, Lowenstine LJ, Gardner MB, Maul DH, Henrickson RV (1984) The pathology of an epizootic of acquired immunodeficiency in rhesus macaques. Am J Pathol 114: 94–103
Petito CK, Navia BA, Cho E-S, Jordan BD, George DC, Price RW (1985) Vacuolar myelopathy pathologically resembling subacute combined degeneration in patients with the acquired immunodeficiency syndrome. N Engl J Med 312: 874–879
Piccardo P, Ceroni M, Rodgers-Johnson P, Mora C, Asher DM, Char G, Gibbs CJ, Gadjusek DC (1988) Pathological and immunological observations on tropical spastic paraparesis in patients from Jamaica. Ann Neurol 73 (Suppl): 156–160
Poiesz BJ, Ruscetti FW, Gazadar AF, Bunn PA, Minna JD, Gallo RC (1980) Detection and isolation of type-C retrovirus particles from fresh and cultured lymphocytes of a patient wth cutaneous T-cell lymphoma. Proc Natl Acad Sci USA 77: 7415–7419
Power MD, Marx PA, Bryant ML, Gardner MB, Barr PJ, Luciw PA (1986) Nucleotide sequence of SRV-1, a type D simian acquired immune deficiency syndrome retrovirus. Science 231: 1567–1572
Price RW, Brew B, Sidtis J, Rosenblum M, Scheck AC, Cleary P (1988) The brain in AIDS: central nervous system HIV-1 infection and AIDS dementia complex. Science 239: 586–592
Rabin H, Benton CV, Tainsky MA, Rice NR, Gilden RV (1979) Isolation and characterization of an endogenous type C virus of rhesus monkeys. Science 204: 841–842
Resnick L, diMarzo-Veronese F, Schupbach J, Tourtellotte WW, Ho DD, Muller F, Shapshak P, Vogt M, Groopman JE, Markham PH, Gallo RC (1985) Intra-blood-brain-barrier synthesis of HTLV-III-specific IgG in patients with neurologic symptoms associated with AIDS or AIDS-related complex. N Engl J Med 313: 1498–1504
Rhodes-Feuillete A, Mahouy G, Lasneret J, Flandrin G, Peries J (1987) Characterization of a human lymphoblastoid cell line permanently modified by simian foamy virus type 10. J Med Primatol 16: 277–289
Ringler DJ, Hunt RD, Desrosiers RC, Daniel MD, Chalifoux LV, King NW (1988) Simian immunodeficiency virus-induced meningoencephalitis: natural history and retrospective study. Ann Neurol 23 (Suppl): S101–S107
Rogers NG, Basnight M, Gibbs CJ, Gajdusek DC (1967) Latent viruses in chimpanzees with experimental kuru. Nature 216: 446–449
Sakakibara I, Sugimoto Y, Sasagawa A, Honjo S, Tsujimoto H, Nakamura H, Hayami M (1986) Spontaneous malignant lymphoma in an African green monkey naturally infected with simain T-lymphotropic virus (STLV) J Med Primatol 15: 311–318
Schneider J, Hunsmann G (1988) Simian lentiviruses—the SIV group. AIDS 2: 1–9
Sharer LR, Baskin GB, Cho E-S, Murphey-Corb M, Blumberg BM, Epstein LG (1988) Comparison of simian immunodeficiency virus and human immunodeficiency virus encephalitides in the immature host. Ann Neurol 23 (Suppl): S108–112
Shaw GM, Harper ME, Hahn BH, Epstein LG, Gajdusek DC, Price RW, Navia BA, Petito CK, O'Hara CJ, Cho E-S, Oleske JM, Wong-Staal F, Gallo RC (1985) HTLV-III infection in brains of children and adults with AIDS encephalopathy. Science 227: 177–182
Sherwin SA, Todaro GJ (1979) A new endogenous primate type C virus isolated from the old world monkey *Colobus polykomos*. Proc Natl Acad Sci USA 76: 5041–5049

Snyder SP, Dungworth DL, Kawakami TG, Callaway E, Lau DT-L (1973) Lymphosarcomas in two gibbons (*Hylobates lar*) with associated C-type virus. J Natl Cancer Inst 51: 89–94
Sonigo P, Barker C, Hunter E, Wain-Hobson S (1986) Nucleotide sequence of Mason-Pfizer monkey virus: an immunosuppressive D-type retrovirus. Cell 45: 375–385
Srivastava BIS, Wong-Staal F, Getchell JP (1986) Human T-cell leukemia virus I provirus and antibodies in a captive gorilla with non-Hodgkin's lymphoma. Cancer Res 46: 4756–4758
Stowell RC, Smith EK, Espana C, Nelson VG (1971) Outbreak of malignant lymphoma in rhesus monkeys. Lab Invest 25: 476–479
Stromberg K, Benveniste RE, Arthur LO, Rabin H, Giddens WE, Ochs HD, Morton WR, Tsai C (1984) Characterization of exogenous type D retrovirus from a fibroma of a macaque with simian AIDS and fibromatosis. Science 224: 289–292
Thayer RM, Power MD, Bryant ML, Gardner MB, Barr PJ, Luciw PA (1987) Sequence relationships of type D retroviruses which cause simian acquired immune deficiency syndrome. Virology 157: 317–329
Theilen GH, Gould D, Fowler M, Dungworth DL (1971) C-type virus in tumor tissue of a Woolly monkey (*Lagothrix* sp.) with fibrosarcomas. J Natl Cancer Inst 47: 881–889
Todaro GJ, Leiber MM, Benveniste RE, Sherr CJ, Gibbs CJ, Gajdusek DC (1975) Infectious primate type C viruses: three isolates belonging to a new subgroup from the brains of normal gibbons. Virology 67: 335–343
Todaro GJ, Sherr CJ, Benveniste RE (1976) Baboons and their close relatives are unusual among primates in their ability to release nondefective endogenous type C viruses. Virology 72: 278–282
Todaro GJ, Benveniste RE, Sherr CJ, Schlom J, Schidlovsky G, Stephenson JR (1978a) Isolation and characterization of a new type D retrovirus from the Asian primate *Presbytis obscurus*. (spectacled langur). Virology 84: 189–194
Todaro GJ, Benveniste RE, Sherwin SA, Sherr CJ (1978b) MAC-1, a new genetically transmitted type C virus of primates: "low frequency" activation from stumptail monkey cell cultures. Cell 13: 775–782
Todaro GJ, Sherr CJ, Sen A, King N, Daniel MD, Fleckenstein B (1978c) Endogenous new world primate type C viruses isolated from an own monkey (*Aotus trivirgatus*) kidney cell line. Proc Natl Acad Sci USA 75: 1004–1008
Tsai C-C, Follis KE, Warner TFCS (1987) Vertical transmission of simian retroviruses. Lab Invest 56: 81A
Tsujimoto H, Seiki M, Nakamura H, Watanabe T, Sakakibara I, Sasagawa A, Honjo S, Hayami M, Yoshida M (1985) Adult T-cell leukemia-like disease in monkey naturally infected with simian retrovirus related to human T-cell leukemia virus type I. Jpn J Cancer Res 76: 911–914
Tsujimoto H, Kodama T, Fukusawa M, Miura T, Speidel S, Nakai M, Cooper RW, Hayami M (1988) Isolation and characterization of simian immunodeficiency virus from mandrill (Abstract). Book II. 4th International conference on AIDS, Stockholm p 98
Vazeux R, Brousse N, Jarry A, Henin D, Marche C, Vedrenne C, Mikol J, Wolfe M, Michon C, Rozenbaum W, Bureau J-F, Montagnier L, Brahic M (1988) AIDS subacute encephalitis: identification of HIV-infected cells. Am J Pathol 126: 403–410
Watanabe T, Seiki M, Tsujimoto H, Miyoshi I, Hayami M, Yoshida M (1985) Sequence homology of the simian retrovirus genome with human T-cell leukemia virus type I. Virology 144: 59–65
Weiss R, Teich N, Varmus H, Coffin J (1982) RNA tumor viruses. Cold Spring Harbor Laboratory, New York, pp 25–207
Wolfe LG, Smith RK, Deinhardt F (1972) Simian sarcoma virus, type I (*Lagothrix*): focus assay and demonstration of nontransforming associated virus. J Natl Cancer Inst 48: 1905–1908
Yamamoto N, Himuma Y, zur Hausen H, Schneider J, Hunsmann G (1983) African green monkeys are infected with adult T-cell leukaemia virus or a closely related agent. Lancet i: 240–241

II. Human Retrovirus Infections of the Nervous System

Human T-Cell Leukemia Virus Type 1
Infections of the Nervous System

Neurological Disorders Associated with HTLV-1

D. E. McFarlin[1] and H. Koprowski[2]

[1] National Institutes of Health, Neuroimmunology Branch, Bldg. 10, Rm. 5B-16, NINCDS, Bethesda, MD 20892, USA
[2] The Wistar Institute, 36th St. at Spruce, Philadelphia, PA 19104, USA

Current Topics in Microbiology and Immunology, Vol. 160

1 Introduction

Retroviruses have been known to be related to a variety of neoplastic, immunodeficient, and chronic degenerative disorders (WEISS et al. 1984; HAASE 1986) for some time, but the 1980s will surely be recorded as the decade of medical history when human disorders related to these agents were recognized. Human T-cell leukemia virus type I (HTLV-I) was first isolated from a patient with cutaneous T-cell lymphoma in 1980 (POIESZ 1980). This led to the widespread belief that similar agents would be related to other human disorders particularly ones affecting the nervous system. In 1985 GESSAIN et al. described antibodies to HTLV-I in patients with tropical spastic paraparesis (TSP), a chronic neurological syndrome which can probably be produced by a number of causes (JOHNSON and MCARTHUR 1987), and KOPROWSKI et al. (1985) detected antibodies against HTLV-I in multiple sclerosis (MS) and proposed that this disease is associated with an HTLV-I-like agent. Subsequently, there has been a burst of activity pursuing the possibility that neurological disorders have a retrovirus etiology. Numerous new findings have been reported, and it is apparent that a progressive disorder primarily, but not exclusively, affecting the spinal cord is linked to HTLV-I; there are also reasons for believing that similar mechanisms may be operative in MS. Currently, there is considerable discussion, as well as debate, about the validity of both old and new findings. Progress is being made, although there is not consensus.

The purpose of this chapter is to review the current status of neurological disorders associated with HTLV-I and related agents. Both the exciting recent developments and the time-honored clinical/pathological concepts of neurological disease are discussed. Because this field is evolving and because international criteria for the diagnosis as well as terminology have not been agreed upon, the developments will be presented historically.

2 Chronic Progressive Myelopathy Syndrome

2.1 Historical Aspects

2.1.1 Tropical Myeloneuropathies

Reports from tropical regions have primarily been descriptive and have documented a number of disorders. The descriptive term "tropical myeloneuropathy"

(TM) has been applied to conditions which vary clinically, pathologically, and etiologically (Dumas 1983; Roman et al. 1985b). Frequently a number of different entities have been grouped under a single descriptive term. This is illustrated by the neurological disorders in Jamaica where a variety of syndromes collectively called "Jamaican neuropathy" have been carefully described over the past century. Over 500 patients with a form of sensory and motor neuritis were reported in 1888 and 1897 (Strachan 1897). In 1918, an "acute outbreak of neuritis" on a sugar plantation was described (Scott 1918). Careful review of the subject (Rodgers-Johnson et al. 1988) disclosed no additional publications until the report of 100 cases of "a neuropathic syndrome of unknown origin" in 1956 (Cruickshank 1956). This condition became known as "Jamaican neuropathy," although the clinical findings in many patients clearly pointed to involvement of the spinal cord. In an extensive review of 206 cases including 11 autopsies (Montgomery et al. 1964), the clinical features, laboratory findings, pathology, and possible etiology were discussed. The patients were divided into two groups on the basis of clinical features, and approximately 15% were placed in an "ataxic group." These patients frequently had sensory abnormalities in the lower extremities and showed posterior column dysfunction producing sensory ataxia, deafness, and retrobulbar neuropathy, the CSF was normal. The second group had a spastic presentation with upper motor neuron abnormalities. A significant percentage of cases had pleocytosis, elevated CSF protein, and a Lang curve which, in view of present knowledge, indicated elevated immunoglobulins. Because of the latter, an infectious etiology was discussed. The histopathology was that of a chronic meningomyelitis with damage to the long tracks.

It now seems likely that a number of different diseases were historically grouped under the general heading "Jamaican neuropathy." Similar sagas describing neurological disorders in other areas have been reviewed (Roman et al. 1985b). On the basis of clinical findings TM is divided into two syndromes, tropical ataxic neuropathy (TAN) and tropical spastic paraplegia (TSP). TAN clinically is manifested by burning feet and other parathesias. Proprioceptive dysfunction is striking and accounts for the ataxia. The ataxic group of patients in Montgomery's report probably would be classified as TAN, and there are descriptions of numerous patients with a similar syndrome in many regions of Africa, Malaysia, and India. Nutritional deficiencies contribute to many cases, but malabsorption and toxins can also be associated with the condition. Although there are probably many causes of the syndrome, there are not convincing data linking TAN to retrovirus (Roman et al. 1985b), except for one patient reported by Rodgers-Johnson et al. (1985).

Tropical spastic paraplegia is primarily manifested by upper motor neuron dysfunction which begins in the lower extremities and is usually progressive. Posterior column dysfunction can occur. The spastic group of patients reported by Montgomery et al. (1964) would probably be diagnosed as TSP. Subsequently, additional cases were reported from Jamaica and similar syndromes described in Africa (Wallace and Cosnett 1983), the Pacific coast of Colombia

(Zaninovic et al. 1981), the Seychelles Islands (Kelly and DeMol 1982), and South India (Mani et al. 1969).

2.1.2 Association of TSP with HTLV-I

As recently as 1985 a number of possible etiologies for TSP were discussed (Roman et al. 1985a). Subsequently, antibodies against HTLV-I in patients with TSP from Martinique were described (Gessain et al. 1985). This was shortly followed by a number of reports from other investigators in Jamaica and Colombia (Rodgers-Johnson et al. 1985) as well as other areas of the Caribbean and the Seychelles Islands. The prevalence of antibodies in affected individuals varied in different reports and it was noted that a higher frequency occurred in family members of affected individuals than in the population in general (Vernant et al. 1987).

2.1.3 Description of HAM

In 1986 Osame reported patients from the Kyushu region of Japan with a progressive myelopathy which was designated HTLV-I-associated myelopathy (HAM) (Osame et al. 1986). These investigators emphasized that the diagnosis of TSP in these patients was inappropriate because Japan does not lie within the tropics. In addition, the presence of antibodies against HTLV-I was considered essential. Consequently, all patients with HAM, by definition, are HTLV-I positive. However, it has recently been called to our attention that some Japanese patients have similar clinical syndromes but lack antibodies to HTLV-I (Itoyomama 1988).

2.1.4 Tropical Spastic Paraplegias Associated with HTLV-I and HAM Are the Same

During 1986 and 1987 a flurry of reports emphasized similarities and differences between TSP associated HTLV-I and HAM. Recently, the features in TSP and HAM were compared by Roman and Osame, who have concluded that the disorders are identical (Roman and Osame 1988). Although this opinion awaits universal acceptance, it is unlikely to be seriously challenged. The term chronic progressive myelopathy (CPM) associated with HTLV-I has been used by others (Brew and Price 1988; Bhagavati et al. 1988), and in the remainder of this chapter CPM will be used. The syndrome is endemic in the tropics and Japan but also occurs in West Indian immigrants to England (Newton et al. 1987) and the United States (Brew and Price 1988; Bhagavati et al. 1988). Sporadic cases have been observed in the United States and in France (Tournier-Lasserve et al. 1987). As discussed below, there is not agreement on the clinical diagnosis of such cases. In addition, it is apparent that not all cases of progressive paraplegia are related to HTLV-I, and other etiologies can produce the clinical syndrome (Johnson and McArthur 1987).

2.2 Clinical Features

2.2.1 Neurological Findings

Clinically, the disorder is confined to the nervous system and typically presents as weakness and stiffness in the lower extremities. Initially, this is usually unilateral but progresses to both legs; the upper extremities and/or the bladder may also be involved. The rate of progression varies. Back pain and sensory disturbances occur in most patients. The age of onset is usually between 20 and 55 years with a mean of approximately 40 years (OSAME et al. 1987; RODGERS-JOHNSON et al. 1988). Symptoms in a 13-year-old male have been reported (ZANINOVIC 1987). In all geographic locations a female/male predominance has been observed (ROMAN and OSAME 1988).

2.2.2 Cerebrospinal Fluid

A mild pleocytosis is common, and elevated CSF immunoglobulins are present in most patients; this is due to local synthesis of CSF immunoglobulin. IgG in the CSF shows an oligoclonal banding pattern after electrophoresis (CERONI et al. 1988; GESSAIN et al. 1988). The oligoclonal IgG bands have been demonstrated to react with components of the HTLV-I virus. IgM and IgA antibodies directed to HTLV-I are also present in the CSF (GUPTA et al. 1988). These findings provide evidence of a local CSF immune response against HTLV-I, which supports its etiological role in the pathogenesis.

2.2.3 Evoked Potentials

Evoked potential studies are now widely used to assess neurological patients. The role of these electrophysiological procedures is to define subclinical lesions and to provide objective evidence of dysfunction in patients with nonspecific symptoms. Abnormalities demonstrated by evoked responses are not pathognomonic for specific disease. These studies detect the presence of lesions, but provide no information about cause (CHIAPPA 1983).

In CPM, lesions are not confined to the spinal cord, and prolonged visual evoked potentials have been observed in approximately 30% of affected individuals. Abnormal auditory evoked potentials have been reported in a small percentage of patients (NEWTON et al. 1987; BHAGAVATI et al. 1988).

2.2.4 Magnetic Resonance Imaging

Magnetic resonance imaging lesions occurred in approximately 50% of patients. These usually occur in the paraventricular regions but have also been observed in the brain and cerebellum. MRI examinations of the spinal cord have mostly been normal, but in a few patients findings consistent with atrophy have been observed (NEWTON et al. 1987; SHEREMATA et al. 1987; BHAGAVATI et al. 1988). Individuals with antibody to HTLV-I but no neurological symptoms can have MRI lesions (MATTSON et al. 1987).

2.3 Neuropathology

Although pathological findings have been described, additional studies, particularly complete examinations of the nervous system using modern techniques, would be valuable. Gross abnormalities are minimal, and it is of particular interest that distinct plaques characteristic of MS have not been described. Microscopically, there is a chronic meningitis with thickening of the pia arachnoid and inflammation. The most extensive changes are in the spinal cord and include perivascular inflammation and degeneration of the long tracts (MONTGOMERY et al. 1964). These show loss of myelin with some preservation of axons (PICCARDO et al. 1988) and reactive astrocytes (JOHNSON et al. 1988). Perivascular inflammation also occurs in the periventricular region and in the cerebellum but is considerably less intense than in the spinal cord (MONTGOMERY et al. 1964; AKIZUKI et al. 1987). Although electron microscopy was performed in one case, the fixation was poor, and virus particles could not be clearly identified (LIBERSKI et al. 1988).

2.4 Antibody to HTLV-I

2.4.1 General

After the report of antibody to HTLV-I in TSP (GESSAIN et al. 1985), similar findings have been confirmed by many investigators. Serological studies have been facilitated and influenced by research in two other areas: (a) the association betweeen HTLV-I and ATL and (b) the AIDS epidemic. Findings have varied among different series which are possibly related to a number of factors including: (a) disease definition, (b) source and type of antigen, (c) methods used for antibody detection, and (d) criteria for a significant reaction.

2.4.2 HTLV-I Viruses and Antigens

Both Hut 102, originally derived from a black American (POIESZ et al. 1980), and MT-2, developed by coculturing blood lymphocytes from an ATL patient with umbilical cord leukocytes (MIYOSHI et al. 1981), as well as antigen preparations derived from these lines, have been used to detect antibody in patients with CPM. Although the genome sequences of Hut 102 and MT-2 are nearly identical, the biological properties of cell lines infected with viruses may vary and affect and antigenic properties. Another potential variable concerns the source of antigen. Some investigators prepared antigen from the above MT-2 (OSAME et al. 1987) or

Hut 102 cell lines (GESSAIN et al. 1985), while others have relied on commercially available antigen preparations.

2.4.3 Assays

Immunofluorescence using HUT 102 or MT-2 cell lines, agglutination of gelatin particles coated with lysates of HTLV-I-infected cell lines, and enzyme-linked immunoadsorbent assay (ELISA) (SAXINGER and GALLO 1983) have been used to screen specimens for antibody reactivity. A radioimmunoassay which uses purified p24 labeled with ^{125}I has also been developed. Most investigators also use Western blot (WB) to confirm positive findings (ALEXANDER et al. 1985). This detects reactivity to individual proteins that have been separated by SDS PAGE and transferred to nitrocellulose. A positive response is determined by reactivity to at least three proteins including the P24. Reactivity to individual proteins has also been detected by a radioimmunoprecipitation assay (RIPA). In this procedure, which is primarily used as a research technique, infected cells are labeled with [^{35}S]methionine or [^{14}C]lysine and lysed with detergents. The radiolabeled lysate is incubated with test sample and the immune complexes absorbed to fixed staphylococci. The eluted antibodies are separated by SDS PAGE and visualized by autoradiography (JACOBSON et al. 1988b).

2.5 Cellular Immunity

2.5.1 Lymphocyte Morphology and Surface Molecules

Large bizarre cells designated as adult T-cell leukemia (ATL)-like have been described in the blood and CSF of many patients with CPM (OSAME et al. 1987; JOHNSON et al. 1988; ROMAN and OSAME 1988). These were identified in stained smears of peripheral blood or CSF. Examination of peripheral blood leukocytes (PBLs) by immunofluorescence and cytofluorography using a combination of forward and right-angle scattering has shown a number of abnormalities. Cells that fall into the lymphocyte-scattering gates show normal distributions of CD3, CD4, and CD8 molecules. However, the lymphocytes from most patients show increased expression of IL-2 receptors as detected by antibody to Tac. Increased expression of HLA DR molecules occurs in approximately 50% of the patients (JACOBSON et al. 1988a). These findings are consistent with lymphocyte activation (GREENE et al. 1986).

Cytofluorographic studies of PBLs have shown an increased number of large cells in the blood of nine patients (JACOBSON et al. 1988a). These are of T-cell lineage and express CD3 molecules. Many of the large cells expressed are Tac^+ and DR^+. This, however, varies, and patients from some regions have large cells that are $CD3^+$ but that lack IL-2 receptors. The expression of CD4 and CD8 molecules on the large $CD3^+$ PBLs also varies and in most patients there is a mixture of such cells with different phenotypes.

2.5.2 Lymphocyte Response

Blood lymphocytes cultured in vitro undergo spontaneous proliferation as assessed by the incorporation of [^{3}H]thymidine (JACOBSON et al. 1988a; JOHNSON et al. 1988). This has been observed in patients from the Caribbean, Peru, and Japan and resembles an effect produced by infecting normal lymphocytes with HTLV-I (GAZZOLO and DODON 1987). This marked degree of spontaneous lymphocyte proliferation obscures the detection of lymphoproliferative responses to specific antigens. As with antibodies to HTLV-I, the occurrence of spontaneous lymphocyte proliferation has been observed in asymptomatic normal individuals in the geographic location endemic for HTLV-I and CPM.

2.5.3 Cytotoxic T-Lymphocytes

The generation of cytotoxic T-lymphocytes (CTLs) against virus-infected targets has been assessed in a few patients with TSP. Measles virus specific CTLs are $CD4^+$ T cells which react with measles antigens in association with class II MHC molecules. This reactivity is reduced in most patients with MS (JACOBSON et al. 1985). Measles virus-specific CTLs have been found to be decreased in a few CPM patients. Generation of CTLs to other viruses is also defective in some CPM patients but is less pronounced than that seen in AIDS (JACOBSON et al. 1988a). Natural killer activity has been observed to be reduced in a few patients (JOHNSON et al. 1988).

2.6 Isolation of HTLV-I

HTLV-I has been isolated from the blood and CSF of patients from Japan (HIROSE et al. 1986; TSUJIMOTO et al. 1988) and tropical regions (DEFREITAS et al. 1987a; JACOBSON et al. 1988b). This was accomplished by either cocultivation of lymphocytes from the blood or CSF with normal lymphocytes or by establishing long-term T-cell lines which were examined for the expression of HTLV-I antigen and genome. T cells are either treated with mitogens (DEFREITAS et al. 1987a) or stimulated with antibody to the CD3/T-cell receptor complex (WEBER et al. 1985) and cultured with Il-2 and feeder cells consisting of irradiated normal blood monocytes. The expression of HTLV-I antigen was detected by immunofluorescence or RIPA. This varied with the passage history and in some cell lines was observed within 1 week, but in others it was not detected until approximately 10 weeks. Budding virions have also been observed by electron microscopy (DEFREITAS et al. 1987a; HIROSE et al. 1986; JACOBSON et al. 1988b), and the presence of infectious virus has been demonstrated by cocultivation.

HTLV-I genome has been detected in cells by in situ hybridization, using a ribo probe, and by Southern blotting using a full-length HTLV-I probe. Using this procedure the DNA from most patients has shown a restriction map consistent with HTLV-I (JACOBSON et al. 1988b; BHAGAVATI et al. 1988).

However, DNA from a few cell lines has shown a different pattern (GREENBERG et al. 1988). Neither HTLV-I antigen nor genome can be detected in T-cell lines from some patients by immunofluorescence and Southern blotting, respectively. However, amplification of genetic sequences by the polymerase chain reaction (PCR) has demonstrated HTLV-I proviral DNA in some cell lines that were negative by the less sensitive techniques (BHAGAVATI et al. 1988).

As pointed out in Sect. 2.3, the neuropathology is characterized by perivenular inflammatory cells, and it seems likely that some of these contain HTLV-I genome; however, this has not been demonstrated and, to date, no isolations from the nervous system have been reported. A possible exception is case 10–13 described by DEFREITAS. Although an HTLV-I agent has been isolated (DEFREITAS et al. 1987a), the diagnosis of this patient is being debated as discussed below.

2.7 Characterization of HTLV-I

As indicated above, the restriction maps of DNA obtained from patients with neurological diseases are consistent with HTLV-I. One complete proviral genome obtained from the CSF of a Japanese patient has been sequenced and shows 97% homology with HTLV-I derived from ATL (TSUJIMOTO et al. 1988). The findings from DNA obtained from a Haitian patient are consistent with this conclusion (REDDY et al. 1988).

Even though existing data indicate that the viruses obtained from patients with neurological diseases are similar to those derived from ATL, the integration of the HTLV-I genome appears to be different. In ATL it is well established that monoclonal integration occurs, but in patients with CPM the findings (YOSHIDA et al. 1987; BHAGAVATI et al. 1988; GREENBERG et al. 1988) indicate polyclonal integration. All of the data had been derived from lymphocyte studies, and information about genome in the nervous system is lacking.

3 Multiple Sclerosis

3.1 Historical Aspects

3.1.1 Clinical Pathological Concepts

Although the pathogenesis and etiology of MS are not known, epidemiological findings indicate an association with an environmental agent, and a possible viral etiology has long been postulated (JOHNSON 1985). As with the TM syndromes, some variation and, at times, confusion concerning the definition of the disease has occurred. However, these have been considerably less because criteria for the clinical diagnosis of MS have been developed and are reassessed periodically.

The early clinical and pathological descriptions of MS were carefully reviewed by DEJONG, and it is generally accepted that there are descriptions of the disease in the medical literature as early as the 1830s (DEJONG et al. 1970). Charchot is given credit for the clinical recognition of the disorder and noting that the neurological abnormalities correlate with *sclerosis en plaques* present at autopsy. It was soon recognized that the disease can have a variety of clinical courses that lead to common pathological features.

3.1.2 Development of Diagnostic Criteria

After MS became defined as a clinical-pathological entity, a number of therapeutic agents were tried but none ameliorated or cured the disease. Assessment of treatment was complicated by heterogeneous clinical features of the disease, and it became apparent that criteria were needed to enable sound experimental trials to be designed in the disease. A committee chaired by Schumacher subsequently developed six criteria that were essential to make a *clinically definite* diagnosis of MS (SCHUMACHER et al. 1965).

In abbreviated form these are:

1. There must be objective abnormalities on neurological examination attributable to CNS dysfunction.
2. Upon neurological examination or by history, there must be involvement of two or more separate areas of the CNS.
3. The objective neurological evidence of CNS disease must reflect predominantly white matter involvement.
4. Involvement of the neuroaxis must have occurred temporally in one of several patterns; either two or more episodes of worsening separated by a period of 1 month or more, or a slow or stepwise progression of signs and symptoms over a period of 6 months.
5. Age of the patient at onset of the disease must fall between 10 and 50 years.
6. Patients' signs and symptoms cannot be explained better by some other disease process.

3.1.3 Use of Laboratory Criteria for Diagnosis

In time other investigators proposed modification of the Schumacher criteria, and subsequently the Medical Advisory Board of the International Federation of MS Societies performed an international survey on the problems relating to diagnosis. Consensus was not obtained, but most responding individuals agreed that some form of criteria should be used (BAUER 1980). During this period, use of laboratory procedures such as examination of the CSF, evoked potentials, and neuroimaging were considered to have diagnostic relevance, and it became common practice of many clinicians to incorporate findings from such studies in the management of patients. In 1983 a new set of diagnostic criteria which incorporated the use of laboratory findings were proposed by a group of neurologists (POSER et al. 1983). Included in the recommendations of this group were new categories designated as "laboratory-supported." According to this proposal it was possible to make a diagnosis of laboratory-supported clinically definite MS on the basis of a single attack with paraclinical evidence of an additional one. It should be emphasized that these criteria were proposed for investigators studying the pathogenesis and treatment of multiple sclerosis. Many neurologists currently use these revised criteria (McDONALD and SILBERBERG 1986).

3.2 Clinical Features

3.2.1 Neurological Findings

The clinical features of MS are well known (HASHIMOTO and PATY 1986; GOODMAN and McFARLIN 1987). Symptoms include motor weakness, incoordination, sensory disturbances, visual abnormalities due to optic neuritis, and difficulties with ocular motility. Bladder abnormalities and sexual dysfunction also occur. The course is quite variable and can take many forms. Most patients with the disease have an exacerbating/remitting course. Others have a chronic progressive process which primarily affects the spinal cord. This can be clinically indistinguishable from CPM associated with HTLV-I as discussed above. An acute form of this disease also occurs, but is uncommon. Some patients present with monosymptomatic syndromes such as optic neuritis or transverse myelitis which, in time, can either evolve into classical MS or persist as focal abnormalities. These produce special problems in diagnosis and management (McDONALD and SILBERBERG 1986).

3.2.2 Cerebrospinal Fluid

Most patients with MS, up to 97% in some series, have abnormalities of CSF immunoglobulins (TOURTELLOTTE and WALSH 1984; PAPADOPOULOS et al. 1987). These include elevated levels of IgG or an increased IgG index, which is due to increased synthesis within the CNS (TOURTELLOTTE and WALSH 1984). IgA and IgM are also elevated in some patients. The CSF IgG of most patients forms oligoclonal bands when electrophoresed. These findings are not specific for MS and can be seen in other disorders such as syphilis and SSPE. However, a major difference is that in MS the antigen reactivity of most CSF immunoglobulins is not known. This is in contrast to other disorders such as SSPE, chronic rubella encephalopathy, and syphilis in which the CSF immunoglobulins react with the causative infectious agent (JOHNSON and NELSON 1977).

3.2.3 Evoked Response Studies

In evaluation of visual-evoked response in MS patients, the amplitude and latency of the major positive peak (P2 or P100) are determined and compared between the two sides. In acute optic neuritis, the latency of the P100 is markedly prolonged, with an initial decrease in amplitude that correlates with reduced visual acuity. With clinical improvement in acuity, the amplitude normalizes, but the latency may be prolonged indefinitely. Thus, the VER latency is often abnormal in patients with normal acuity. Abnormalities are found in 75% of patients with clinically definite MS.

Brain stem auditory-evoked responses are a sensitive tool for evaluation of auditory pathways and nearby structures. One representative study showed abnormality in 57% of patients with clinically defined brain stem involvement, and 21% in those with no clinical signs. Short-latency somatosensory-evoked responses are generally elicited from median, ulnar, peroneal, and tibial nerves and are abnormal in about 80% of patients with clinically definite MS. The greatest yield is derived from stimulation of nerves in the lower extremities. Abnormalities are more frequent if there are sensory disturbances of pyramidal signs in the leg being studied (Chiappa 1983).

3.2.4 Neuroimaging and MRI

Multiple sclerosis was one of the first diseases studied using magnetic resonance imaging (MRI). Findings can vary depending upon the techniques employed. The procedure is particularly useful in detecting small plaques in the brain stem, cerebellum, and spinal cord. The findings in a recently completed research study demonstrate the power of this procedure (Ormerod et al. 1987). Lesions were detected in 113/114 patients with clinically definite disease. All but one of these showed lesions in the periventricular regions; the remaining patient had a solitary lesion in the brain stem. Sagittal images of the cervical spinal cord showed lesions in 20/24 patients. A significant percentage of patients presenting with isolated syndromes including the spinal cord, optic nerves, or brain stem showed MRI evidence of disseminated lesions.

It is clear that MRI is the imaging technique of choice for the diagnosis of MS. However, positive MRI scans without appropriate clinical evidence are nothing more than suggestive of the diagnosis. Further, a positive study only indicates the presence of a lesion but provides no information about etiology. Periventricular lesions in white matter similar to those seen in MS have been found in patients with vascular diseases including infarctions, systemic lupus, erythematosis, Sjogren's syndrome, other forms of vasculitis, and inflammatory conditions such as Behçet's disease and encephalitis (Ormerod et al. 1987). Also, MRI lesions in the white matter have been attributed to aging and are thought to have a vascular basis (Brant-Zawadski et al. 1985).

3.3 Neuropathology

The classical neuropathological lesion in MS is a plaque of demyelination (PRINEAS 1985). These are large enough to be visualized grossly and occur throughout the white matter. However, there are areas of predilection including the optic nerves and chiasm, periventricular regions, corticomedullary junction, subpial regions of the cervical cortex and brain stem, and certain regions of the spinal cord. Plaques vary in size, and coalescence of small lesions is believed to produce larger ones. Microscopic examination shows that the lesions occur in a perivenular distribution and have myelin loss wth preservation of axons, nerve cells, and blood vessels. Perivascular cuffs of lymphocytes and plasma cells are common in acute and subacute lesions. In older lesions cellular reactions are less, and there is astrogliosis that gives rise to the sclerotic texture. The mechanisms responsible for myelin degeneration are not known, but the disorder has been viewed as a disease of oligodendroglia (LUMSDEN 1971) which is either infected by a latent virus or the target of an immune-mediated process. In recent years the concept has emerged that the disease is due to destruction of the myelin sheath which is believed to be mediated by macrophages (PRINEAS 1985).

3.4 Immunological Studies

3.4.1 Studies of Antibody

Considerable effort has been devoted to seeking an antigen or agent which reacts with immunoglobulins synthesized in the CNS of patients with MS. Such studies have been unsuccessful. Beginning with ADAMS and IMAGAWA (1962), there have been many reports of increased antibody production to measles virus in MS. This is well documented, but, in addition, evidence of intrathecal antibody synthesis to a number of other agents has been reported (JOHNSON et al. 1985) and many patients show increased antibody production to more than one agent (SALMI et al. 1983). More recently, antibodies reactive with the p24 *gag* protein of HTLV-1 have been described in 30%–40% of blood and CSF from MS patients. Although these findings were contested (MADDEN et al. 1988; RICE et al. 1986; GESSAIN et al. 1986) by some workers who used different assays, the observation may represent the first clue to a possible role of human retrovirus in MS (see below). Even when demonstrable antibody reactivity is taken into consideration, it is only possible to account for a small amount of the total reactivity of the immunoglobulins that are produced in the CNS. This has led to the concept that in MS the overactivity of antibody-producing cells is related to a defect in immunoregulation (REDER and ARNANSON 1985).

3.4.2 Studies of Cellular Immune Function

Evidence of abnormal cellular immunity to autoantigens and viruses has also been sought. In spite of considerable efforts, a specific response to neural antigens has not been consistently observed (REDER and ARNANSON 1985; JOHNSON et al. 1985). However, observations from several laboratories indicate that the cellular immune response to measles virus is reduced in MS patients. This is manifested by reduced lymphokine production (UTTERMOHLEN and ZABRISKIE 1973), decreased lymphoproliferation (MCFARLAND et al. 1979), and impaired capacity to generate measles virus-specific CTLs (JACOBSON et al. 1985). It is of interest that measles virus-specific CTLs are $CD4^+$ T cells that recognize the virus in connection with class II MHC molecules.

3.5 Pathogenesis and Etiology

3.5.1 General Concepts

Neither the pathogenesis nor the etiology of MS has been established, but both environmental and genetic components are believed to contribute. Support for a genetic contribution has been derived from studies of the disease in families and particularly in twins. A higher concordance in monozygotic than dizygotic twins has been observed, and many clinically normal monozygotic twins of individuals with MS show CSF and/or neuroimaging findings consistent with subclinical disease (MCFARLAND et al. 1985; EBERS et al. 1986). The findings suggest that more than one genetic influence is operative. This has led to the hypothesis that genes encoding both the MHC and the T-cell receptor contribute to the pathogenesis (GOODMAN and MCFARLIN 1987).

Support for an environmental factor comes from epidemiological studies which have identified areas of high risk and low risk. Investigation of migrants between areas of different risks indicates a relationship between the disease and exposure to an environmental agent before adolescence (KURTZKE 1985). Additional evidence for an environmental agent has come from the identification of clusters and epidemics of the disease. The most convincing was the occurrence of an epidemic of MS on the Faroe islands which followed the distribution of British troops during World War II (KURTZKE and HYLLESTED 1986). Although both environmental and genetic factors are believed to be operative in MS, the pathogenetic mechanisms responsible for production of the disease process are at present not known. An immune-mediated process has long been suspected, and there are reasons for believing that immune regulation is abnormal in the disease. The details of this concept have been reviewed elsewhere (REDER and ARNANSON 1985).

3.5.2 Viral Etiology

The possibility that MS is related to an infectious agent was suggested in 1884 (MARIE 1884) but has not been proven in spite of extensive investigations. There are at least two major lines of indirect evidence supporting a relationship between MS and a virus (JOHNSON 1985). The first is epidemiological data which indicate an environmental agent is involved. Secondly, a number of demyelinating disorders in animals and humans have been shown to be related to viruses. Over the past 4 decades, at least 12 different agents have been proposed, but none has been confirmed. Currently, serious consideration is being given to the role of HTLV-I or a related retrovirus in the etiology of the disease.

3.5.3 Multiple Sclerosis As a Retroviral Disease

A retroviral etiology for MS first proposed by KOPROWSKI in 1985 (KOPROWSKI et al. 1985) would explain many of the puzzling features of the disorder. Retroviruses characteristically produce persistent infections of monocytes and T-lymphocytes (WONG-STAAL and SALLO 1985). This could lead to both immunoregulatory abnormalities and chronic disease. Infected monocytes could transfer the virus to the CNS, as is believed to occur in AIDS and visna. Recent observations in visna, a naturally occurring lentivirus infection of sheep, may provide insight into the pathogenesis of MS. The neurological disease is characterized by demyelination and perivascular inflammation, and, although monocytes containing virus genome are found in CNS, these are rare, about $1/10^4$ (HAASE 1986). An immunopathological reaction with release of lymphokines may produce the demyelinating process (KENNEDY et al. 1985).

3.6 Data Supporting Retrovirus in MS

3.6.1 Overlap Between MS and CPM

In the recent literature are descriptions of patients who either have been diagnosed as MS or who would meet some criteria for this diagnosis and also have high antibodies to HTLV-I. Approaches and opinions differ. For example, case II reported by TOURNIER-LASSERVE et al. (1987) was concluded to have TSP because of the high titers to HTLV-I. The details of this case have bearing on the issue. The white patient resided in Paris and had never traveled to a geographic region endemic for HTLV-I. The clinical features indicated involvement of the spinal cord, and MRI showed periventricular white matter lesions. "MS was the first diagnosis suspected," but it was concluded that the patient had an HTLV-I-associated spastic paraparesis because of high antibody titers to HTLV-I in the serum and CSF.

Another example is patient 10–13 studies by DEFREITAS et al. (1987a). This Cuban male presented with clinical features and MRI abnormalities consistent with MS; at autopsy both granulomatous vasculitis and demyelinating lesions of different ages were seen. This patient had no antibodies to HTLV-I but viral antigen was expressed by cultured CSF cells and in brain tissue. Furthermore, ultrastructural examination of cultures of CSF cells showed particles compatible in size with a retrovirus.

The above cases are relevant to diagnostic considerations for patients with CPM. Many individuals with this disorder have paraclinical evidence of lesions that indicate the disease is not confined to the spinal cord. It is noteworthy that in 1964 approximately 15% of patients with the spastic form of "Jamaican neuropathy" were concluded to have retrobulbar neuritis (MONTGOMERY et al. 1964). In view of this, it has been pointed out that patients with CPM would meet the recent criteria for a diagnosis of laboratory-supported MS (POSER et al. 1983). Opponents of this view emphasize the sixth Schumacher criterion, i.e., a disease process that does not have a better explanation. Points of debate are whether the association with HTLV-I offers a better explanation and whether in the future reclarification of this disease in relation to HTLV-I infection will be indicated, particularly if the role of HTLV-I in the pathogenesis of the diseases becomes *more firmly* established.

3.6.2 Antibody Studies

As mentioned above, in the studies reported by KOPROWSKI et al. (1985) antibody to purified p24 of HTLV-I was detected by ELISA. This was found both in patients from Florida who might have been exposed to HTLV-I and patients from Sweden, which is not endemic for this agent. Subsequently, reactivity for *gag* proteins was detected in approximately 25% of Japanese MS patients using Western blotting (OHTA et al. 1986) and in 40% of patients using "particle agglutination" test with p24 antigen by SAIDA (DEFREITAS et al. 1987b). Other investigations (RICE et al. 1986; MADDEN et al. 1988) have been unable to confirm the presence of antibody in the sera and CSF from MS patients using ELISA. As emphasized in Sect. 2.4 and by DEFREITAS et al. (1987b), techniques vary in the capacity to detect antibody. However, it should be pointed out that even in studies that describe antibody reactivity to HTLV-I in MS, it has not been detected in all patients, and, when present, the amount is relatively low in comparison to the CPM syndrome. This suggests that if a retrovirus is related to MS, the agent has some cross-reactivity with HTLV-I but is not identical.

3.6.3 Detection of HTLV-I Proteins and Genome

HTLV-I antigen identified by the APAAP test with anti-*gag* antibody was found in cells originating from blood and/or CSF of two MS patients and maintained in culture (DEFREITAS et al. 1989). It is interesting that although HTLV-I antibodies in serum and CSF of one of these patients reacted with all proteins of HTLV-I, the *gag* antigen expressed by the infected cells seemed to differ antigenically from the prototype HTLV-I expressed by the Hut 102 cells. This suggested that the retrovirus may not be identical with HTLV-I prototype.

HTLV-I-related nucleotide sequences were detected in CSF lymphocytes by in situ hybridization with a 3′ RNA probe which primarily contained *LTR* region. These were infrequent (0.001% to 0.01%) and found under low stringency; however, identical studies of control cultures were negative (KOPROWSKI et al. 1985). It should be pointed out that the conditions are different from those used to detect HTLV-I genome in lymphocytes cultured from a patient with CPM.

Genome-containing cells were easily demonstrated under high stringency. In a subsequent study, small numbers of small cells (10^{-4}–10^{-5}) that were labeled with an HTLV-I cDNA probe were found in the blood of MS patients, but positive cells were also detected at approximately the same frequency in normal controls. The same percentage of cells in MS patients and controls were reactive with the pBR322 probe lacking HTLV-I sequences. Because of this background, these findings are difficult to interpret and point out the need for additional studies with specific probes. In the same study, HTLV-I genome was not detected in MS lesions by in situ hybridization with the same cDNA probe. As pointed out in Sect. 2.6 above, HTLV-I antigen and genome were not detectable in some T-cell lines derived from CPM patients. However, proviral DNA was identified in such cells after amplification by PCR. Similar studies are now being conducted with T-cell lines derived from MS patients and it will be of great interest to follow the findings that are certain to be forthcoming.

4 Comments

Over the past 4 years, research in a number of laboratories has established an association between chronic human neurological diseases and HTLV-I or closely related agents. These accomplishments are exciting, but it is clear that considerably more effort will be required to understand the pathogenesis of the diseases mentioned above, as well as other disorders which may be shown to be associated with human retroviruses. The immunological and molecular biological procedures currently being used to study these viruses are both powerful and highly sensitive. It is apparent that in the near future a vast amount of new data will emerge. It is essential to emphasize that the interpretation of this information will require that parallel studies be conducted in normals and patients with a variety of neurological disorders.

It should also be recognized that demonstration of associations between viruses and neurological diseases, although highly important, do not define pathogenetic mechanism. For example, the studies conducted to date indicate that many individuals become exposed and even persistently infected with HTLV-I but do not develop neurological diseases. More research on the possible effect of these agents on the nervous system as well as immunological and immunopathological mechanisms is clearly needed.

References

Adams JM, Imagawa DT (1962) Measles antibodies in multiple sclerosis. Proc Soc Exp Biol Med 111: 562–566

Akizuki S, Nakazato O, Higuchi Y, Tanabe K, Setoguchi M, Yoshida S, Miyazaki Y, Yamamoto S, Sudou S, Sannomiya K, Okajima T (1987) Necropsy findings in HTLV-I associated myelopathy. Lancet i: 987: 156–157

Alexander SS, Tai CC, Ting RL et al. (1985) Utilization of the Western blot assay for immunodeficiency disease. Proc Am Soc Microbiol 24: 297
Bauer HJ (1980) Clinical definition and assessment of the course of MS. IMAB-Enquete concerning the diagnostic criteria for MS. In: Bauer HJ, Poser S, Ritter G (eds) Progress in multiple sclerosis research. Springer Berlin Heidelberg New York, pp 553–563
Bhagavati S, Ehrlich G, Kula RW, Kwok S, Sninsky J, Udani V, Poiesz BJ (1988) Detection of human T-cell lymphoma/leukemia virus type I DNA and antigen in spinal fluid and blood of patients with chronic progressive myelopathy. N Engl J Med 318: 1141–1147
Brant-Zawadski M, Fein G, Van Dyke C, Kierman R, Davenport L, de Groot J (1985) MR imaging of the aging brain: patchy white matter lesions and dementia. Am J Neutoradiol 6: 675–682
Brew BJ, Price RW (1988) Another retroviral disease of the nervous system: chronic progressive myelopathy due to HTLV-I. N Engl J Med 318: 1195–1198
Bunn PA, Schechter GP, Jaffe E et al. (1983) Clinical course of retrovirus-associated adult T-cell lymphoma in the United States. Engl J Med 309: 257–264
Ceroni M, Piccardo P, Rodgers-Johnson P, Mora C, Asher DM, Gajdusek DC, Gibbs CJ (1988) Intrathecal synthesis of IgG antibodies to HTLV-I supports an etiological role for HTLV-I in tropical spastic paraparesis. Ann Neurol S23: S188–S191
Chiappa EH (1983) Evoked potentials in clinical medicine. Raven, New York
Cruickshank EK (1956) A neuropathic syndrome of uncertain origin. West Indian Med J 5: 147–158
DeFreitas E, Wroblewska Z, Maul G, Sheremata W, Ferrante P, Lavi E, Harper M, Di Marzo-Veronese F, Koprowski H (1987a) HTLV-I infections of cerebrospinal fluid T cells from patients with chronic neurologic disease. Aids Res Hum Retrovirus 3: 19–32
DeFreitas E, Saida T, Iwaski Y, Koprowski H (1987b) Association of human T-lymphocytotrophic viruses in chronic neurological disease. Ann Neurol 21: 215–216
DeFreitas E et al (1989) Expression of a retrovirus partially related to HTLV-I in T cells from an American patient with MS. In: Roman G et al (eds) HTLV-I and the nervous system. Alan R. Liss, New York, pp 407–420
DeJong RN (1970) Multiple sclerosis, history, definition and general considerations. In: Vinken PJ, Bruyn G (eds) Handbook of clinical neurology. Multiple sclerosis and other demyelinating diseases, vol 9. Elsevier Science, Amsterdam, pp 45–62
Dumas M (1983) Neuro-myelopathies tropicales. Rev Neurobiol 29: 155–194
Ebers GC, Bulman DE, Sadovnick AD, Paty DW, Warren S, Hader W, Murray TJ, Seland TP, Duquette P, Grey T, Nelson R, Nicolle M, Brunet D (1986) A population-based study of multiple sclerosis in twins. N Engl J Med 315: 1638–1642
Gazzolo L, Dodon MD (1987) Direct activation of resting T lymphocytes by human T-lymphotropic virus type I. Nature 326: 714–717
Gessain A, Barin F, Vernant JC, Gout O, Maurs L, Calender A, de-The G (1985) Antibodies to human T lymphotrophic virus type I in patients with tropical spastic paraparesis. Lancet ii: 407–410
Gessain A, Vernant JC et al. (1986) Lack of HTLV-I and LAV/HTLV-III antibodies in patients with multiple sclerosis in France and French West Indies. Br Med J 293: 424–425
Gessain A, Caudie C, Gout O, Vernant JC, Maurs L, Giordano C, Malone G, Tournier-Lasserve E, Essex M, de-The G (1988) Intrathecal synthesis of antibodies to human T lymphotrophic virus type I and the presence of IgG oligoclonal bands in the cerebrospinal fluid of patients with endemic tropical spastic paraparesis. J Infect Dis 157: 1226–1234
Goodman A, McFarlin DE (1987) Multiple sclerosis. Curr Neurol 7: 91–128
Greenberg SJ, Jacobson S, Waldmann TA, McFarlin DE (1988) Molecular analysis of HTLV-I proviral integration and T cell receptor beta-chain gene arrangement in tropical spastic paraparesis. J Inf Dis (to be published)
Greene WC, Leonard WJ, Depper JM, Nelson DL, Waldman TA (1986) The human interleukin-2 receptor: normal and abnormal expression in T cells in leukemias induced by the human T-lymphotropic retroviruses. Ann Intern Med 105: 560–572
Gupta A, Mingioli ES, Mora C, McFarlin DE (1988) Detection of anti-IgG, IgM, and IgA antibodies against HTLV-I in sera and CSF of patients with tropical spastic paraparesis (TSP). Neurology 38: 89
Haase AT (1986) Pathogenesis of lentivirus infections. Nature 322: 130–136
Hashimoto SA, Paty DW (1986) Multiple sclerosis. DM 32(9): 518–589
Hauser S, Aubert C, Burks JS, Kerr C, Lyon-Caen O, de-The G, Brahic M (1986) Analysis of human T-lymphotropic virus sequences in multiple sclerosis tissue. Nature 322: 176–177

Hirose S, Uemura Y, Fujishita M, Kitagawa T, Yamashita M, Imamura J, Uhtsuki Y, Taguchi H, Miyoshi I (1986) Isolation of HTLV-I from cerebrospinal fluid of a patient with myelopathy. Lancet ii: 397–398

Itoyomama Y (1988) Presented at symposium on retroviruses and chronic neurological diseases. Fondation Marcel Merieux. Les Pensieres. Veyrier-du-lac. Annecy, France

Jacobson S, Flerlage M, McFarland HF (1985) Impaired measles virus-specific cytotoxic T cell responses in multiple sclerosis. J Exp Med 162: 839–850

Jacobson S, Zaninovic V, Mora C, Rodgers-Johnson P, Sheremata W, Gibbs CJ, Gajdusek CD, McFarlin DE (1988a) Immunological findings in neurological diseases associated with antibodies to HTLV-I: activated lymphocytes in progressive spastic paraparesis. Ann Neurol 23: S196–S200

Jacobson S, Raine CS, Mingioli ES, McFarlin DE (1988b) Isolation of an HTLV– I like retrovirus from patients with tropical spastic paraparesis (TSP) by activation with antibodies to the CD3 complex. Nature 331: 540–543

Johnson KP, Nelson BJ (1977) Multiple sclerosis: diagnostic usefulness of cerebrospinal fluid. Ann Neurol 2: 425–431

Johnson RT (1985) Viral aspects of multiple sclerosis. In: Vinken PJ, Bruyn GW, Klawans HL, Koetsier JC (eds) Handbook of clinical neurology. Demyelinating diseases, vol 11. Elsevier Science, Amsterdam, pp 319–336 (47 revised series 3)

Johnson RT, McArthur JC (1987) Myelopathies and retroviral infections (Editorial). Ann Neurol 21: 113

Johnson RT, Griffin DE, Arregui A, Mora C, Gibbs CJ, Cuba JM, Trelles L, Vaisberg A (1988) Spastic paraparesis and HTLV– I infection in Peru. Ann Neurol 23: S151–155

Kelly R, De Mol B (1982) Paraplegia in the islands of the Indian Ocean. Afr J Neurol Sci 1: 5–7

Kennedy PGE, Narayan O, Ghotbi Z, Hopkins J, Gendelman HE, Clements JE (1985) Persistent expression of Ia antigen and viral genome in visna-maedi virus induced inflammatory cells. J Exp Med 162: 1970–1982

Koprowski H, DeFreitas EC, Harper ME, Sandberg-Wollheim M, Sheremata WA, Robert-Guroff M, Saxinger CW, Feinberg MB, Wong-Staal F, Gallo RC (1985) Multiple sclerosis and human T-cell lymphotrophic retroviruses. Nature 318: 154–160

Kurtzke JF (1985) Epidemiology of multiple sclerosis. In: Vinken PJ, Bruyn GW, Klawans HL, Koetsier JC (eds) Handbook of clinical neurology. Demyelinating diseases, vol 9. Elsevier Science, Amsterdam, pp 259–287 (47 revised series 3)

Kurtzke JF, Hyllested K (1986) Multiple sclerosis in the Faroe Islands. II. Clinical update, transmission and the nature of MS. Neurol 36: 307–328

Liberski PP, Rodgers-Johnson P, Char G, Piccardo P, Gibbs CJ, Gajdusek DC (1988) HTLV-I-like viral particles in spinal cord cells in Jamaican tropical spastic paraparesis. Ann Neurol 23 (Suppl): S185–S187

Lumsden CE (1971) The immunogenesis of the multiple sclerosis plaque. Brain Res 28: 365–390

Madden DL, Mundon FK, Fuccillo DA, Calabrese V, Roman GC, Sever JL (1988) Antibody to human and simian retrovirus, HTLV-I, HTLV-II. HIV, STLV-III, and SRV-I not increased in patients with multiple sclerosis. Ann Neurol 23: S171–S173

Mani KS, Mani AJ, Montgomery RD (1969) A spastic paraplegic syndrome in South India. J Neurol Sci 9: 179–199

Marie P (1884) Sclerose en plaques et maladies infectieuses. Prog Med 12: 287–289

Mattson DH, McFarlin DE, Mora C, Zaninovic V (1987) Central nervous system lesions detected by magnetic resonance imaging in an HTLV-I antibody positive symptomless individual. Lancet ii: 49

Mattson DH, Gupta A, Jacobson S, Mora C, Rodgers-Johnson P, Morgan OStc, Zaninovic V, McFarlin DE (1988) Clinical and immunological investigations of tropical spastic paraparesis. Neurology 38 (Suppl 1): 166

McDonald WI, Silberberg DH (1986) Diagnosis of multiple sclerosis. In: McDonald WI, Silberberg DH (eds) Multiple sclerosis. Butterworths, London, pp 1–10

McFarland HF, McFarlin DE (1979) Cellular immune response to measles, mumps, and vaccinia viruses in MS. Ann Neurol 6: 101–106

McFarland HF, Greenstein JI, McFarlin DE, Eldridge R, Xu X, Krebs H (1985) Family and twin studies in multiple sclerosis. Ann NY Acad Sci 436: 118–124

Miyoshi I, Kubonishi I, Sumida M, Hiraki S, Tsubota T, Kimura I, Miyamoto K, Sato J (1980) A novel T-cell line derived from adult T-cell leukemia. Jpn J Cancer Res 71: 155–156

Miyoshi I, Kubonishi I, Yoshimoto S, Shiraishi Y (1981) A T-cell line derived from normal human cord leukocytes by coculturing with human leukemic T-cells. Jpn J Cancer Res 72: 978–981

Montgomery RD, Cruickshank EK, Robertson WB, McMenemey WH (1964) Clinical and pathological observations on Jamaican neuropathy. A report on 206 cases. Brain 87(3): 425–462

Newton M, Miller D, Rudge P, Cruickshank K, Dalgleish A, Clayden S, Moseley I (1987) Antibody to human T-lymphotropic virus type 1 in West-Indian Born Uk residents with spastic paraparesis. Lancet i: 415–416

Ohta M, Ohta K, Mori F, Nishitani H, Saida T (1986) Sera from patients with multiple sclerosis react with human T-cell lymphotrophic virus *gag* proteins but not *env* proteins-Western blotting analysis. J Immunol 137(11): 3440–3443

Ormerod IEC, Miller DH, McDonald WI, Du Boulay EPGH, Rudge P, Kendell BE, Moseley IF, Johnson G, Tofts PS, Halliday AM, Bronstein AM, Scaravilli F, Harding AE, Barnes D, Zilkha KJ (1987) The role of NMR imaging in the assessment of multiple sclerosis and isolated neurological lesions. A quantitative study. Brain 110(6):1579–1616

Osame M, Usuku K, Izumo S, Ijichi N, Amitani H, Igata A, Matsumoto M, Tara M (1986) HTLV-I associated myelopathy: a new clinical entity. Lancet i: 1031–1032

Osame M, Igata A, Matsumoto M, Usuku K, Kitajima I, Takahashi K (1987a) On the discovery of a new clinical entity: human T-cell lymphotrophic virus type I-associated myelopathy (HAM). Adv Neurol Sci Tokyo Jpn 31: 727–745

Osame M, Matsumoto M, Usuku K, Izumo S, Ijichi N, Amitani H, Tara M, Igata A (1987b) Chronic progressive myelopathy associated with elevated antibodies to human T-lymphotrophic virus type I and adult T-cell leukemia-like cells. Ann Neurol 21: 117–122

Papadopoulos NM, McFarlin DE, Patronas NJ, McFarland HF, Costello R (1987) A comparison between chemical analysis and magnetic resonance imaging with the clinical diagnosis of multiple sclerosis. Am J Clin Pathol 88: 365–368

Piccardo P, Ceroni M, Rodgers-Johnson P, Mora C, Asher DM, Char G, Gibbs CJ, Gajdusek DC (1988) Pathological and immunological observations on tropical spastic paraparesis in patients from Jamaica. Ann Neurol 23(Suppl): S156–S160

Poiesz BJ, Ruscetti FW, Gazdar AF, Bunn PA, Minna JD, Gallo RC (1980) Detection and isolation of type C retrovirus particles from fresh and cultured lymphocytes of a patient with cutaneous T-cell lymphoma. Proc Natl Acad Sci USA 77: 7415–7419

Poser CM, Paty DW, Scheinberg L, McDonald I, Davis FA, Ebers GC, Johnson KP, Sibley WA, Silberberg DH, Tourtellotte WW (1983) New diagnostic criteria for multiple sclerosis: guidelines for research protocols. Ann Neurol 13: 227–231

Prineas JW (1985) The neuropathology of multiple sclerosis. In: Vinken PJ, Bruyn GW, Klawans HL, Koetsier JC (eds) Handbook of clinical neurology. Demyelinating diseases, vol 8. Elsevier Science, Amsterdam, pp 213–257 (47 revised series 3)

Reddy EP, Mettus RV, DeFreitas E, Wroblewska Z, Cisco M, Koprowski H (1988) Molecular cloning of human T-cell lymphotrophic virus type I-like proviral genome from the peripheral lymphocyte DNA of a patient with chronic neurologic disorders. Proc Natl Acad Sci USA 85: 3599–3603

Reder AT, Arnanson BGW (1985) Immunology of multiple sclerosis. In: Vinken PJ, Bruyn GW, Klawans HL, Koetsier JC (eds) Handbook of clinical neurology. Demyelinating diseases, vol 12. Elsevier Science, Amsterdam, pp 337– 395 (47 revised series 3)

Rice GPA, Armstrong HA, Bulman DE, Paty DW, Ebers GC (1986) Absence of antibody to HTLV– I and II in sera of Canadian patients with multiple sclerosis and chronic myelopathy. Ann Neurol 20: 533– 534

Rodgers-Johnson P, Gajdusek DC, Morgan OStC, Zaninovic V, Sarin PS, Graham DS (1985) HTLV-I and HTLV-III antibodies and tropical spastic paraparesis. Lancet ii: 1247–1248

Rodgers-Johnson P, Morgan OStC, Mora C, Sarin P, Ceroni M, Piccardo P, Garruto RM, Gibbs CJ, Gajdusek DC (1988) The role of HTLV-I in tropical spastic paraparesis in Jamaica. Ann Neurol 23: S121–126

Roman GC, Osame M (1988) Identity of HTLV-I-associated tropical spastic paraparesis and HTLV-I-associated myelopathy. Lancet i: 651

Roman GC, Spencer PS, Path MRC, Schoenberg BS (1985a) Tropical myeloneuropathies: the hidden endemias. Neurology 35: 1158–1170

Roman GC, Roman LN, Spencer PS, Schoenberg BS (1985b) Tropical sapstic paraparesis: a neuroepidemiological study in Columbia. Ann Neurol 17: 361–365

Salmi A, Reunanen M, Ilonen J, Panelius M (1983) Intrathecal antibody synthesis to virus antigens in multiple sclerosis. Clin Exp Immunol 52: 241–249

Saxinger C, Gallo RC (1983) Application of the indirect enzyme-linked immunoabsorbent assay microtest to the detection and surveillance of human T-cell leukemia-lymphoma virus. Lab Invest 49: 371–377

Schumacher GA, Beebe G, Kibler RF et al. (1965) Problems of experimental trials of therapy in multiple sclerosis: report by the panel of the evaluation of experimental trials of therapy in multiple sclerosis. Ann NY Acad Sci 122: 552

Scott HH (1918) Investigation into an acute outbreak of 'central neuritis.' Ann Trop Med Parasitol 12: 109–196

Sheremata WA, Quencer R, Gatti E, DeFreitas E, Harper M, Koprowski H (1987) Magnetic resonance imaging (MRI) of tropical spastic paraplegia. Neurology 37 (Suppl 1): 322

Strachan H (1897) On a form of multiple sclerosis prevalent in the West Indies. Practitioner 59: 477–484

Tournier-Lasserve E, Gout O, Gessain A, Iba-Zizen MT, Lyon-Caen O, Lhermitte F, de The G (1987) HTLV-I brain abnormalities on magnetic resonance imaging and relation with multiple sclerosis. Lancet ii: 49–50

Tourtellotte WW, Walsh MJ (1984) Cerebrospinal fluid profile in multiple sclerosis. In: Poser CM (ed) The diagnosis of multiple sclerosis. Thieme, New York, pp 165–178

Tsujimoto T, Teruuchi T, Imamura J, Shimotohno K, Miyoshi I, Miwa M (1988) Nucleotide analysis of a provirus derived from HTLV-I-associated myelopathy (HAM)

Uttermohlen V, Zabriskie JB (1973) A suppression of cellular immunity in patients with multiple sclerosis. J Exp Med 138: 1591–1596

Vernant JC, Maurs L, Gessain A, Barin F, Gout O, Delaporte JM, Sanhadji K, Buisson G, de The G (1987) Endemic tropical spastic paraparesis associated with human T-lymphotropic virus type I: a clinical and seroepidemiological study of 25 cases. Ann Neurol 21: 123–130

Wallace ID, Cosnett JE (1983) Unexplained spastic paraplegia. S Afr Med J 63: 689–691

Weber WE, Buurman WA, Marc M, Vandermeeren PP, Raus JCM (1985) Activation through CD3 molecule leads to clonal expansion of all human peripheral blood T lymphocytes: functional analysis of clonally expanded cells. J Immunol 135: 2337–2342

Weiss R, Teich N, Varmus H, Coffin J (eds) (1984) Origins of contemporary RNA tumor virus research, 2nd edn. Cold Spring Harbor Laboratory, New York, pp 1–24 (Molecular biology of tumor viruses: RNA tumor viruses 1)

Wong-Staal F, Gallo RC (1985) Human T-lymphotropic retroviruses. Nature 317: 395–403

Yosida M, Osame M, Usuku M, Matsumoto M, Igata A (1987) Viruses detected in HTLV-I associated myelopathy and adult T-cell leukemia are identical on DNA blotting. Lancet i: 1085–1086

Zaninovic V (1987) Tropical spastic paraparesis. Lancetii: 280

Zaninovic V, Biojo R, Barreto P (1981) Paraparesia espastica del Pacifio. Columbia Med 12: 111–117

An HTLV-I Transgenic Mouse Model: Role of the Tax Gene in Pathogenesis in Multiple Organ Systems

M. I. NERENBERG

1 Introduction

The human T-lymphotropic virus type I (HTLV-I) has been associated with a lymphoproliferative disease called adult T-cell leukemia/lymphoma (ATLL) (POIESZ et al. 1980; YOSHIDA et al. 1982; HINUMA et al. 1981; KALYANARAMAN et al. 1982) and a neurodegenerative disease called tropical spastic paraparesis (TSP) (GESSAIN et al. 1985; RODGERS-JOHNSON et al. 1985; BARTHOLOMEW et al. 1986) or HTLV-I-associated myelopathy (HAM) (OSAME et al. 1986a). These diseases appear to share few common clinical features. ATLL has a prolonged latency, but once established is frequently rapidly fatal (BRODER et al. 1984). The disturbance consists of a monoclonal expansion of mature peripheral T cells each of which contains copies of retroviral sequences (POIESZ et al. 1980; MIYOSHI et al. 1980; YAMAMOTO et al. 1981; POPOVIC et al. 1983; CHEN et al. 1983; YOSHIDA et al. 1984). Death results from consequences of uncontrolled cellular proliferation and dysfunction. In contrast, the associated neurologic disease usually shows a shorter latency period and a slowly progressive course (ROMAN 1988; OSAME et al. 1986a). Though variable degrees of HTLV-I-infected lymphocytic infiltration into the CNS may be seen in this disease, the cellular target for viral infection within the nervous system remains unclear (JACOBSON et al. 1988). To study

Department of Immunology, Scripps Clinic and Research Foundation, 10666 N. Torrey Pines Road, La Jolla, California 92037, USA

Current Topics in Microbiology and Immunology, Vol. 160

potential mechanisms of these disparate HTLV-I-induced diseases, we created a transgenic mouse model in which the *tax* gene (*trans*-activator of transcription) was introduced under control of the HTLV-I viral long-terminal repeat (LTR) promoter.

2 Background HTLV-I

2.1 Biologic Effects of HTLV-I

The consequences of HTLV-I infection have been most carefully studied for T-lymphocytes. Proliferation may be induced by two mechanisms: viral products have an immediate polyclonal mitogenic effect on T cells which is independent of infection. This phenomenon is transient and has recently been postulated to be a mechanism whereby HTLV-I may modify the characteristics of other simultaneously infecting organisms (ZACK et al. 1988). The more profound effect, however, is a monoclonal expansion of mature $CD4^+$-circulating peripheral T cells containing viral sequences. These cells may contain a complete proviral genome and produce infectious virus or contain deleted genomes. Mapping studies of these defective genomes reveals a preferential retention of the 3′ portion, a region which encodes viral regulatory proteins (HIRAMATSU and YOSHIKURA 1986). Infected cells have impaired function and have a limited life span in vitro. However, when exposed to immature human hematopoietic cells, they fuse with and transform T cells by passage of infectious virus. By direct cell-to-cell fusion, these cells may also infect a wide variety of other cell types including fibroblasts, B cells, muscle, and endothelial and osteoid cells (HOXIE et al. 1984; HO et al. 1984; NAGY et al. 1983). Since the virus is carried in patients by widely circulating peripheral T cells, and there appears to be little restriction in cell types which the virus can infect, a wide variety of cell types would be expected to be infected in patients.

2.2 Effects of HTLV-I Regulatory Proteins

Within the 3′ region of the provirus, three nonstructural proteins, p40, p27, and p21, are encoded by a doubly spliced subgenomic mRNA (KIYOKAWA et al. 1985; NAGASHIMA et al. 1986). p27 and p21 are encoded in the same reading frame and are identical except that p27 has an amino-terminal extension. p27 is a phosphoprotein, is found in the nucleus, and appears to modulate viral transactivation through posttranscriptional effects on some of the virus-encoded mRNAs. It has been suggested that levels of this protein may control latency or switching from early to late functions in the virus (ROSENBLATT et al. 1988). The effects of p27 and p21 on endogenous cellular genes remain unclear. The *tax* gene product is the other protein encoded by this mRNA. It is a 40-kd protein which localizes to the nucleus in tissue culture lines (HINRICHS et al. 1987; GOH et al. 1985; SLAMON et al. 1985). Previous in vitro data demonstrated the *tax* gene product to be capable of augmenting transcription of the HTLV-I viral LTR, as well as a number of endogenous cellular genes involved in the control of growth or in the elaboration of cytokines which are likely to have systemic effects. These include the interleukin-2 gene, the interleukin-2 receptor gene (GREENE et al. 1986; MARUYAMA et al. 1987), and the granulocyte-monocyte colony-stimulating factor gene (CHAN et al. 1986). We therefore chose to focus on the *tax* gene as a likely modulator of biologic effects.

3 Transgenic Studies

3.1 Expression Vector

Within the viral LTR U3 region are the *tax*-responsive sequences and viral promoter. Therefore by placing the *tax* gene downstream of the *U3* and *R* regions of the viral LTR, *tax* protein expression is controlled by the viral LTR and is subject to *tax*-mediated autoregulation. Cell-specific expression then reflects the extent to which *tax* and the LTR contribute to viral tropism, though species differences between mouse and man may introduce additional complexities. The construct also contained a 3′ copy of the viral LTR which includes the normal polyadenylation signal. This was placed in a pSV-based vector system which includes the SV40 small t antigen intron as well as an additional polyadenylation site 3′ of the coding region. This construct was shown by in vitro transfection to encode a 40-kd protein capable of transcriptionally transactivating the viral LTR, which was linked to the bacterial chloramphenicol acetyl transferase reporter gene. The *tax*-encoding transcription cassette was purified from flanking bacterial plasmid sequences and microinjected into fertilized mouse oocytes.

3.2 Generation of Transgenic Mice

For the original lines of transgenic *tax* mice, the CD1 mouse strain was chosen as an oocyte donor, as well as breeding partner. Though these are outbred mice and thus are not well defined genetically, they have an unusually low incidence of spontaneous tumors and have a low number of endogenous retroviruses. Subsequent breeding into other strains has shown that the development of phenotypes is not strain dependent. Out of approximately 20 founder mice, *tax* gene expression was assayed on 8, each of which expressed authentic *tax* product. Three of these founder mice were used to establish transgenic lines. The tissue

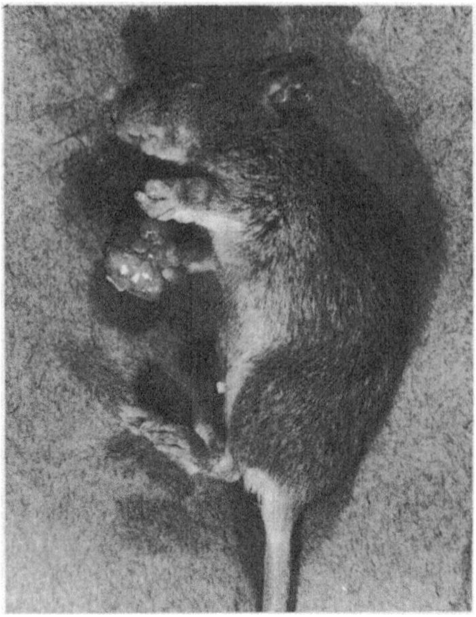

Fig. 1. Multiple simultaneous nerve-associated tumours involving the extremities, ears, nose, and tail. Approximate age of mouse is 4 months.

pattern of *tax* expression was similar in each of these although the overall levels of *tax* protein varied by five- to tenfold between high- and low-level-expressing animals. By Western blot, expression was detected in tail, skeletal muscle, peripheral nerve, and salivary gland. These mice appeared grossly normal until approximately 3 months of age when they developed nodular growths in the tails, ears, extremities, and occasionally around the face (Fig. 1). Tumors were frequently clustered around sites of repeated trauma. Autopsies also revealed tumors at the base of the cranium, which occurred most frequently in pregnant female mice.

3.3 Microscopic Analysis

Histologic analysis showed that the tumors of the extremities and intracranial tumors were associated with peripheral nerve sheaths. No tumors of the CNS were found. Intraductal tumors of the salivary gland were also seen, which were not of nerve origin. The peripheral nerve-associated tumors resulted in gross deformities with a phenotypic resemblance to peripheral form of neurofibromatosis of humans (von Recklinghausen disease). However, histologically they were shown to be composed of perineurial cells and not of Schwann cells. Schwann cells are the predominant proliferating cell type in most cases of human von Recklinghausen disease (ERLANDSON 1985). The tumors showed other unusual features. Perineurial tumors were heavily infiltrated by mature polymorphonuclear monocytes in the absence of tumor necrosis. Regions of peripheral nerves which expressed *tax*, but were morphologically normal, were free of granulocytes. Thus *tax* expression in perineurial tissue alone could not account for PMN infiltration.

3.4 Immunocytochemical Studies

Immunohistochemistry was performed in collaboration with Dr. CLAYTON WILEY, UCSD Department of Pathology, to delineate the sites of expression at a cellular level and further define pathogenesis (NERENBERG and WILEY 1989). On a cellular basis, *tax* expression was high in skeletal muscle, though only a fraction of fibers within a muscle group expressed the protein. This fraction was seen to vary between different anatomic groups of muscles. In addition, all fibers in which expression could be detected exhibited mild to profound atrophy, while the nonexpressing counterparts appeared histologically normal. Histochemical staining for mitochondria and for the enzyme NADH revealed that muscle fibers which expressed *tax* protein and developed atrophy were composed exclusively of oxidative fibers. ATPase staining demonstrated that both types 1 and 2a oxidative fibers were involved. Subcellular localization of *tax* revealed accumulation only in the cytoplasm of muscle fibers.

The pattern of expression of *tax* protein in peripheral nerve was strikingly different. In nerve regions not containing tumors, expression was restricted to perineurial cells, in which both nuclear and cytoplasmic accumulation could be seen. Tumor cells uniformly expressed *tax*, in a predominantly nuclear pattern. Infiltrating granulocytes did not express *tax*. A similar pattern was seen in salivary gland tissue in which only ductal cells expressed and gave rise to intraductal tumors.

4 Implications

These studies establish that a single viral protein may induce pleiomorphic and seemingly opposite pathologic effects dependent on the tissue of expression. This suggests a basis for the disparate diseases of proliferation and atrophy associated with HTLV-I infection in humans. Apparent differences in tissue tropism between human infections and the mouse model may relate to species differences or the ease with which phenotypes are detected in the two systems. The transgenic system is likely to be a far more sensitive assay of the effects of a single gene product, since a copy of this gene is placed into every cell of the animal and has an opportunity to interact with many pathways in the course of animal development. It may thus uncover subtle effects of the virus which are not easily appreciated in human populations. It is interesting to note, for example, that skeletal muscle disease has been associated with HTLV-I infection but not well studied (MORA et al. 1988). It is also a system in which the importance of other potentially important cofactors may be tested. Thus other HTLV-I-encoded factors may be introduced into the germline of mice and interactive affects tested by mating the two strains. Transgenic mice may also be useful in assessing the contributing effects of other infectious agents. For example, ATLL is a disease with a prolonged latency, and which other infections cofactors have been implicated. The reproducible onset and clinical features of diseases in mice also offer an opportunity to analyze molecular pathways contributing to pathogenesis. Future studies using this model will address the contribution of each of other HTLV-I and heterologous virus factors in an attempt to understand the determination of tropism in the CNS and the molecular pathology of TSP.

Acknowledgments. This is Pub. No. 5452-IMM from the Department of Immunology, Scripps Clinic and Research Foundation. This work was supported in part by USPHS training grant AG-00080. I would like to thank Mrs. Gay Schilling for assistance in preparation of this manuscript.

References

Bartholomew CF, Cleghaorn F, Charles W, Ratan P, Roberts L, Maharaj K, Janke N, Daisley H, Hanchard B, Blattner W (1986) HTLV-I and tropical spastic paraparesis. Lancet ii: 99–100
Broder S, Jaffe ES, Blattner W, Gallo RC, Wong-Staal F, Waldman TA, DeVita VT (1984) T cell lymphoproliferative syndrome associated with human T cell leukemia/lymphoma virus. Ann Intern Me 100: 543–557
Cann AJ, Rosenblatt JD, Wachsman W, Shaw NP, Chen ISY (1985) Identification of the gene responsible for human T cell leukemia virus transcriptional regulation. Nature 318: 571–574
Chan JY, Slamon DJ, Nimer SD, Golde DW, Gasson JC (1986) Regulation of expression of human granulocyte/macrophage colony-stimulating factor. Proc Natl Acad Sci USA 83: 8669–8673
Chen IS, Quan SG, Golde DW (1983) Human T cell leukemia virus type II transforms normal human lymphocytes. Proc Natl Acad Sci USA 80: 7006–7009
Erlandson RA (1985) Peripheral nerve sheath tumors. Ultrastruct Pathol 9: 113–122
Gessain A, Vernant JC, Maurs L, Barin F, Gout O, Calender A, DeThe G (1985) Antibodies to human T lymphotropic virus type I in patients with tropical spastic paraparesis. Lancet ii: 407–409
Goh WC, Sodroski J, Rosen C, HaseltineWA (1985) Subcellular localization of the product of the long open reading frame of human T cell leukemia virus type I. Science 227: 1227–1228
Greene WC, Leonard WJ, Wano Y, Svetlik PB, Peffer NK, Sodroski JG, Rosen CA, Goh WC, Haseltine WC (1986) *Trans*-activator gene of HTLV-II induces IL-2 receptor and IL-2 cellular gene expression. Science 232: 877–880
Hinrichs SH, Nerenberg M, Reynolds RK, Khoury G, Jay G (1987) A transgenic mouse model for human neurofibromatosis. Science 237: 1340–1343
Hinuma Y, Nagata K, Hanaoka M, Nakai M, Matsumoto T, Kinoshita K-I, Shirakawa S, Miyoshi I (1981) Adult T cell leukemia: antigen in an ATL cell line and detection of antibodies to the antigen in human sera. Proc Natl Acad Sci USA 78: 6476–6480
Hiramatsu K, Yoshikura H (1986) Frequent partial deletion of human adult T cell leukemia virus type I proviruses in experimental transmission: pattern and possible implication. J Virol 58: 508–512
Ho D, Rota T, Hirsch M (1984) Infection of human endothelial cells by human T lymphotropic virus type I. Proc Natl Acad Sci USA 81: 7588–7590

Hoxie JA, Matthews DM, Cines DB (1984) Infection of human endothelial cells by human T cell leukemia virus type I. Proc Natl Acad Sci USA 81: 7591–7595

Jacobson S, Raine CS, Mingioli ES, McFarlin DE (1988) Isolation of an HTLV-I-like retrovirus from patients with tropical spastic paraparesis. Nature 331: 540–543

Kalyanaraman VS, Samgadharan MG, Nakao Y, Ito Y, Aoki T, Gallo RC (1982) Natural antibodies to the structural core protein (p24) of the human T cell leukemia (lymphoma) retrovirus found in sera of leukemia patients in Japan. Proc Natl Acad Sci USA 79: 1653–1657

Kiyokawa T, Seiki M, Iwashita S, Imagawa K, Shimizu F, Yoshida M (1985) $p27^{x-III}$ and $p21^{X-III}$ Proteins encoded by the pX sequence of human T cell leukemia virus type I. Proc Natl Acad Sci USA 82: 8359–8363

Maruyama M, Shibuya H, Harada H, Hatakeyama M, Seiki M, Fujita T, Inoue J, Yoshida M, Taniguchi T (1987) Evidence for aberrant activation of the interleukin-2 autocrine loop by HTLV-I encoded p40 and T3/Ti complex triggering. Cell 48: 343–350

Miyoshi I, Kubonishi I, Yoshimoto S, Akagi T, Ohtsuki Y, Shiraishi Y, Nagata K, Hinuma Y (1981) Type C virus particles in a cord T cell line derived by co-cultivating normal human cord lymphocytes and human leukaemic T cells. Nature 294: 770–771

Mora CA, Garruto RM, Brown P, Guiroy D, Morgan OSC, Rodgers-Johnson P, Ceroni M, Yanagihara R, Goldfarb LG, Gibbs CJ, Gajdusek DC (1988) Seroprevalence of antibodies to HTLV-I in patients with chronic neurological disorders other than tropical spastic paraparesis. Ann Neurol 23 (Suppl): S192–S195

Nagashima K, Yoshida M, Seiki M (1986) A single species of pX mRNA of human T cell leukemia virus type I encodes *trans*- activator $p40^x$ and two other phosphoproteins. J Virol 60: 394–399

Nagy K, Clapham P, Cheingsong-Popov R, Weiss R (1983) Human T cell leukemia virus type I: induction of syncytia by patients sera. Int J Cancer 32: 321–328

Nerenberg MI, Hinrichs SH, Reynolds RK, Khoury G, Jay G (1987) The *tat* gene of human T lymphotropic virus type 1 induces mesenchymal tumors in transgenic mice. Science 237: 1324–1329

Nerenberg MI, Wiley CA (1989) The HTLV-I *tax* protein induces tissue specific disorders of neoplasia and atrophy in transgenic mice. Am. J. Pathol., in press.

Osame MK, Usuku K, Izumo S, Ijichi N, Amitani H, Igata A, Matsumoto M, Tara M (1986a) HTLV-I associated myelopathy, a new clinical entity. Lancet i: 1932–1033

Osame M, Izumo S, Igata A, Matsumoto M, Matsumoto T, Sonoda S, Tara M, Shibata Y (1986b) Blood transfusion and HTLV-I associated myelopathy. Lancet ii: 104–105

Poiesz BJ, Ruscetti FW, Gadzar AF, Bunn PA, Minna JD, Gallo RC (1980) Detection and isolation of type C retrovirus particles from fresh and cultured lymphocytes of a patient with cutaneous T cell lymphoma. Proc Natl Acad Sci USA 77: 7415–7419

Poiesz BJ, Ruscetti FW, Reitz MS, Kalyanaraman VS, Gallo RC (1981) Isolation of a new type C retrovirus (HTLV) in primary uncultured cells of a patient with Sezary T cell leukaemia. Nature 294: 268–271

Popovic M, Sarin PS, Robert-Guroff M, Kalyanaraman VS, Mann D, Minowada J, Gallo RC (1983) Isolation and transmission of human retrovirus (human T cell leukemia virus). Science 219: 856–859

Robert-Guroff M, Nakao Y, Notake K, Ito Y, Sliski A, Gallo RC (1982) Natural antibodies to human retrovirus HTLV in a cluster of Japanese patients with adult T cell leukemia. Science 215: 975–978

Rodgers-Johnson P, Morgan C, St O, Zaninovic V, Sarin P, Graham DS (1985) HTLV-1 and HTLV-II antibodies and tropical spastic paraparesis. Lancet ii: 1247–1248

Roman GC (1988) The neuroepidemiology of tropical spastic paraparesis. Ann Neurol 23: S113–S120

Rosenblatt JD, Cann, AJ, Slamon DJ, Smalberg IS, Shaw NP, Fujii J, Wachsman W, Chen ISY (1988) HTLV-II transactivation is regulated by the overlapping *tax/rex* nonstructural genes. Science 240: 916–918

Slamon DL, Press MF, Souza CM, Murdock DC, Kline MJ, Golde DW, Gasson JC, Chen ISY (1985) Studies of the putative transforming protein of the type I human T cell leukemia virus. Science 228: 1427–1430

Weiss R (1982) In: Weiss R, Teich N, Varmus H, Coffin J (eds) RNA tumor viruses. Cold Spring Harbor Laboratory, New York, pp 1205–1281

Yamamoto N, Okada M, Koyanagi Y, Kannagi M, Hinuma Y (1981) Transformation of human leukocytes by cocultivation with an adult T cell leukemia virus producer cell line. Science 217: 737–739

Yoshida B, Miyoshi I, Hinuma Y (1982) Isolation and characterization of retrovirus from cell lines of human adult T cell leukemia and its implication in the disease. Proc Natl Acad Sci USA 79: 2031–2035
Yoshida M, Seiki M, Yamaguchi K, Takatsuki K (1984) Monoclonal integration of human T cell leukemia provirus in all primary tumors of adult T cell leukemia suggests causative role of human T cell leukemia virus in the disease. Proc Natl Acad Sci USA 81: 2534–2537
Zack JA, Cann AJ, Lugo JP, Chen ISY (1988) HIV-production from infected peripheral T cells after HTLV-1 induced mitogenic stimulation. Science 240: 1029–1032

Human Immunodeficiency Virus Infections of the Nervous System

Using Synthetic Peptide Reagents to Distinguish Infections Caused by Different HIV Strains

J. W. GNANN, Jr. and M. B. A. OLDSTONE

1 Introduction

As the nucleic acid sequences of additional isolates of human immunodeficiency virus (HIV) have been determined, the remarkable heterogeneity within this group of viruses has become apparent. HIV has been subclassified into HIV type 1, the retrovirus first shown to cause the acquired immunodeficiency syndrome (AIDS), and HIV type 2, a more recently discovered retrovirus prevalent in West Africa. HIV-1 and HIV-2 have identical amino acids at about 59% of positions in the relatively conserved core proteins and at about 42% of positions in the envelope proteins (GUYADER et al. 1987). There is partial antigenic cross-reactivity between the core proteins of the two viruses, but cross-reactivity between envelope proteins has not been described (CLAVEL et al. 1986, 1987; BRUN-VEZINET et al. 1987). Consequently, an enzyme-linked immunosorbent assay (ELISA) containing HIV-1 whole-virus lysate as an antigen detects some HIV-2-positive sera, but is an inconsistent and unreliable test for HIV-2 diagnosis.

Department of Immunology, Scripps Clinic and Research Foundation, 10666 N. Torrey Pines Road, La Jolla, CA 92037, USA

Current Topics in Microbiology and Immunology, Vol. 160

HIV-1 isolates from persons of different geographic areas have significant genetic variability, especially in the *env* gene that encodes the envelope glycoproteins (COFFIN 1986; BENN et al. 1985; HAHN et al. 1985). This genetic variability has not, however, resulted in noticeable antigenic variability as measured by standard ELISA or Western blotting, indicating that important antigenic regions are conserved among these HIV-1 strains (ALIZON et al. 1986; BRUN-VEZINET et al. 1984).

In vitro studies have clearly demonstrated that genetic variations in HIV-1 strains can result in altered biologic properties such as the ability to replicate in particular cell lines (CHENG-MAYER et al. 1987; EVANS et al. 1987; DAHL et al. 1987) or susceptibility to neutralization by immune sera (PRINCE et al. 1987). How strain variation influences biologic properties of the virus in vivo is the subject of ongoing investigation (CHENG-MAYER et al. 1988). An area of particular importance is the effect of genetic variation on neurotropism and neurovirulence (see J. LEVY, this volume).

Efforts to understand the biologic effects of HIV genetic variations in vivo would be bolstered by the availability of serologic tests that could supplant restriction enzyme analysis as the primary technique for identifying infection by particular viral strains. In this chapter, we describe the development and use of synthetic peptide reagents for serologically separating HIV-1 from HIV-2 infections and suggest an approach for distinguishing among various strains of HIV-1

2 Uses of Synthetic Peptide Reagents for HIV Serodiagnosis

Synthetic peptides may be used as immunogens (to generate antibodies that will cross-react with the native protein from which the peptide was derived) or as antigens (to detect humoral or cell-mediated immune responses generated against the native protein). For the serodiagnosis of HIV infection, synthetic peptide antigens that specifically bind anti-HIV immunoglobulins are needed. When compared with antigens prepared from native proteins, such synthetic peptide antigens have the advantages of availability in large quantities, high purity, defined specificity, and lack of infective potential.

3 Selection of Viral Protein Regions for Synthesis

A given viral protein usually contains multiple antigenic determinants. However, many of the antigenic determinants appear to be conformational rather than sequential (SUTCLIFFE et al. 1983). That is, the determinant is assembled from discontinuous segments of the amino acid chain by tertiary folding of the native protein and does not arise from a continuous sequence of amino acids. The challenge for the investigator, therefore, is to select for synthesis those linear amino acid sequences, usually 10–20 residues in length, that have the highest probability of functioning as B-cell antigens.

A number of "rules" have been proposed to help the investigator select the most favorable protein regions for synthesis, although exceptions to these rules are frequent (LERNER 1984). To bind antibody, an antigenic determinant on a native protein must be accessible, so is likely to be exposed on the surface of the protein molecule. Hence, the rules are designed to select those sequences that are presumably located at the protein's surface rather than buried in the interior. Characteristics that may identify B-cell determinants include high local average hydrophilicity (HOPP and WOODS 1981), positions near protein-chain termini (WALTER et al. 1980), high turn potential (SUTCLIFFE et al. 1983; CHOU and FASMAN 1978), and high segmental mobility (WESTHOF et al. 1984; TAINER et al. 1984). In addition, a peptide to to be used as an antigen in a diagnostic immunoassay should be selected from a region of the protein that is conserved among various strains of the virus. This is particularly important with HIV, which contains several hypervariable regions, especially in the external glycoprotein (COFFIN 1986). Computer programs have been devised that use these parameters to predict possible antigenic determinants of HIV-1 proteins (MODROW et al. 1987; PAULETTI et al. 1985; STERNBERG et al. 1987).

4 Antigenic Epitopes of HIV-1

Investigators have synthesized peptides representing computer-predicted epitopes and tested the peptides for reactivity with sera from HIV-1-infected individuals by ELISA or radioimmunoassay (RIA). Two antigenic envelope epitopes have been identified by this approach. Both epitopes are located in hydrophilic regions that are highly conserved among HIV-1 isolates (MODROW et al. 1987). One antigenic site (*env* amino acids 504–518 of strain HTLV-III_B) is located at the carboxy terminus of gp120 (the external glycoprotein) adjacent to the proteolytic cleavage site between gp120 and gp41 (PALKER et al. 1987). A synthetic peptide derived from this region adsorbed a high percentage of the anti-gp120 antibody activity from reactive human sera, indicating that this is an important epitope. However, only about 45% of HIV-1-infected individuals had detectable antibody against this peptide. A second antigenic site (*env* amino acids 735–752 of strain HTLV-III_B) has been described near the carboxy end of gp41 (the transmembrane glycoprotein). A synthetic peptide analogous to this region reacted with a small number of HIV-1-positive sera when used as an antigen in an ELISA (KENNEDY et al. 1986).

Another gp120 determinant with antigenic properties has recently been described (PALKER et al. 1988). This epitope (*env* amino acids 303–321 of strain

HTLV-III$_B$) is in a strongly hydrophilic region with high predicted values for β-turn, flexibility, and surface probability (MODROW et al. 1987). However, the region also exhibits a substantial degree of amino acid variability among HIV-1 strains. A 20-amino-acid synthetic peptide corresponding to the sequence of strain HTLV-III$_B$ reacted in RIA with sera from 21% of HIV-1-seropositive patients. In addition, antibody raised against this peptide had potent strain-specific virus-neutralizing activity, indicating that the site may be an important type-specific neutralizing epitope (PALKER et al. 1988).

In an effort to identify other antigenic sites, we synthesized 15 peptides representing potential epitopes from proteins encoded by the HIV-1 *gag*, *pol*, and *env* genes (GNANN et al. 1987a). Sequences were selected primarily on the basis of hydrophilicity and sequence conservation. Of these 15 peptides, the only one that reacted with significant numbers of sera from HIV-1-infected patients was a 26-amino-acid peptide derived from gp41 (amino acids 584–609 of strain LAV-1$_{BRU}$). When used as an antigen in an ELISA, this peptide-bound serum antibody from 53/53 HIV-1-infected patients and from 0/50-uninfected controls. Other investigators have also synthesized peptides representing various portions of *env* amino acid sequence 584–618 and found significant reactivity with sera from large numbers of HIV-1-seropositive individuals (WANG et al. 1986; SMITH et al. 1987). Interestingly, this highly immunoreactive region was not identified by the predictive computer programs because of its overall hydrophobicity.

5 Immunodominant gp41 Epitope of HIV-1

5.1 Mapping the Binding Site

To better localize the antigenic epitope, we synthesized additional shorter peptides extending across the region of interest from *env* amino acid 584 to 618 (GNANN et al. 1987a). A peptide (*env* amino acids 603–614) that overlapped the carboxy end of the 26 amino acid peptide reacted with 60% of the HIV-1-positive test sera. Another peptide (*env* amino acids 609–620) flanked the carboxy end of the 26-amino-acid peptide and was recognized by 35% of the test sera. However, we also synthesized a 12-amino-acid peptide (*env* amino acids 598–609; LGLWGCSGKLIC) that fully retained the antigenicity of the 26-amino-acid peptide. This 12-amino-acid peptide reacted with 162/163 sera from HIV-1-positive patients in the United States (including asymptomatic individuals, patients with AIDS-related complex, and patients with AIDS) and with 0/19 seronegative control sera (including sera from healthy homosexual men) (GNANN et al. 1987c).

To determine the precise residues essential for antibody binding, a nested set of nine peptides was synthesized with sequential amino acid deletions from the termini of the reactive 12-amino-acid peptide (GNANN et al. 1987b). Stepwise

Table 1. Fine-mapping of the essential structure of an immunodominant determinant of gp41. [Reproduced with permission from GNANN et al. (1987b)]

Amino acid position[a]	Peptide sequence	No. positive/ No. tested[b]	% Positive
A. aa598–609	L G L W G C S G K L I C	22/22	100%
aa599–609	G L W G C S G K L I C	21/22	95%
aa600–609	L W G C S G K L I C	20/22	91%
aa601–609	W G C S G K L I C	19/22	86%
aa602–609	G C S G K L I C	14/22	64%
aa603–609	C S G K L I C	21/44	48%
aa604–609	S G K L I C	0/22	0%
aa600–607	L W G C S G K L	0/22	0%
aa598–603	L G L W G C	0/22	0%
B. aa598–609	L G L W G C S G K L I C	2/22	100%
aa598–609	L G L W G C S G K L I C	2/22	9%

[a]Amino acid positions based on the sequence of LAV-1_{BRU}
[b]Number of HIV-1-positive sera tested by ELISA. All peptides were nonreactive with normal control sera

elimination of amino-terminal residues caused incremental reductions in reactivity of the peptide with HIV-1-positive sera (Table 1A). The smallest peptide that retained some antigenic activity was a seven-amino-acid sequence (603–609) with a cysteine residue at each end that reacted with 48% of test sera. Removal of the cysteine residue from either end eliminated antigenicity, suggesting that disulfide bond formation might be important in the antigenic conformation of the peptide. To test this hypothesis further, a peptide was synthesized with a serine substituted for cysteine at position 609 (Table 1B). This substituted peptide reacted with only 9% of HIV-1-positive sera, thus confirming the importance of the cysteine residues for antigenicity (GNANN et al. 1987b).

5.2 Kinetics of Antibody Response to the gp41 Epitope

The appearance of IgM and IgG antibodies was measured with a peptide ELISA in 13 serum specimens collected from 4 patients before and after seroconversion to HIV-1 (SCHRIER et al. 1988). Low titers of IgM antibody were detected early during the course of infection, then waned or disappeared in subsequent specimens. In contrast, the IgG antibody response to this epitope was higher in titer and persisted.

5.3 Antibody Response to gp41 Amino Acids 598–609 Is T-cell Dependent

To investigate the T-cell dependence of the antibody response to the gp41 epitope, homozygous nude (athymic) mice and their heterozygous (nude/+)

littermates were immunized with the 12-amino-acid peptide (SCHRIER et al. 1988). The four immunocompetent mice responded with high levels of peptide-specific IgG at 7 and 14 days postimmunization, but the four T-cell-deficient nude/nude mice failed to do so (Fig. 1). Both groups of mice made low levels of peptide-specific IgM, detected on day 7 only. These data indicate that the murine IgG response to the gp41 epitope is T-cell dependent whereas the IgM response is T-cell independent.

5.4 Reactivity of Sera from African Patients with Synthetic HIV-1 Peptides

The 12-amino-acid gp41 peptide exhibited excellent sensitivity and specificity when used to test sera from HIV-1-infected patients in the United States. To extend this observation to sera from Africans, we tested specimens from Zairian AIDS patients. The peptide ELISA was positive with only 33 of 38 HIV-1-positive sera from Zaire (and 0/21 seronegative controls), a level of sensitivity significantly lower than that seen with sera from American AIDS patients (GNANN et al. 1987c).

The region of the transmembrane glycoprotein from which the 12-amino-acid peptide was derived is relatively conserved among the strains of HIV-1 that have

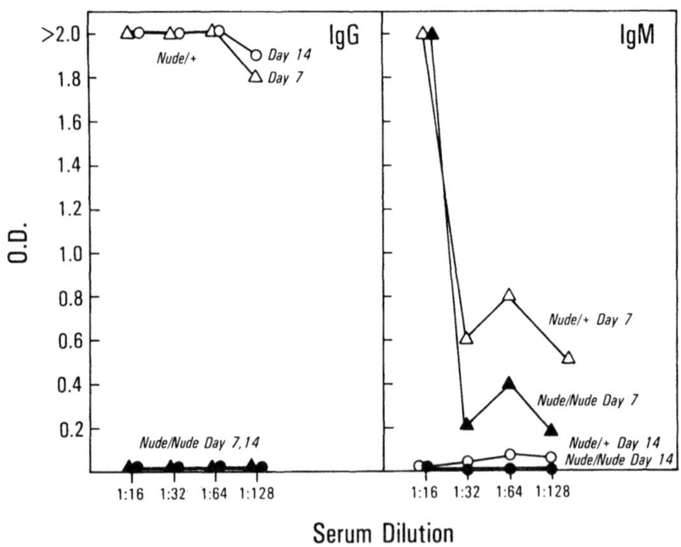

Fig. 1. The IgG response to the gp41 epitope (amino acids 598–609) is T-cell dependent. Nude/nude athymic mice (*closed symbols*) and nude/+ immunocompetent mice (*open symbols*) were immunized with 100 μg peptide in incomplete Freund's adjuvant. Sera obtained 7 (Δ) and 14 (O) days postimmunization were tested by ELISA for peptide-specific IgM and IgG. O.D., optical density at 492 nm. [Reproduced with permission from SCHRIER et al. (1988)]

Table 2. Amino acid conservation in the transmembrane glycoproteins of HIV-1 and HIV-2. (Adapted with permission of GNANN et al. (1989)

Isolate	Source	Sequence[a,b,c]	Reference
		598 609	
HIV-1 (LAV-1_{BRU})	France	*Leu*GlyIle*TrpGlyCys*SerGlyLysLeuIle*Cys*	WAIN-HOBSON et al. (1985)
HIV-1 (HTLV-III_{B})	United States	------------------------------------	RATNER et al. (1985)
HIV-1 (ARV-2)	United States	------------------------------------	SANCHEZ-PESCADOR et al. (1985)
HIV-1 (HTLV-III_{MN})	United States	------------------------------------	GURGO et al. (1988)
HIV-1 (HTLV-III_{SC})	United States	------------------------------------	GURGO et al. (1988)
HIV-1 (WMJ-1)	Haiti	------------------------------------	STARCICH et al. (1986)
HIV-1 (HTLV-III_{RF})	Haiti	------------------------------------	STARCICH et al. (1986)
HIV-1 (Zr6)	Zaire	------------------------------------	SRINIVASAN et al. (1987)
HIV-1 (LAV-1_{MAL})	Zaire	------------------------His --------	ALIZON et al. (1986)
HIV-1 (LAV-1_{ELI})	Zaire	----Met ------------His --------	ALIZON et al. (1986)
HIV-1 (Z3)	Zaire	----Leu------------------------------	WILLEY et al. (1986)
HIV-1 (NY5)	United States	----Met -------------------------	WILLEY et al. (1986)
HIV-1 (CDC-451)	United States	----Phe------------------------------	DESAI et al. (1986)
		592 603	
HIV-2 (HIV-2_{ROD})	Cape Verde Islands	*Leu*AsnSer*TrpGlyCys*AlaPheArgGlnVal*Cys*	GUYADER et al. (1987)

[a]Amino acid position numbers for HIV-1 are based on the sequence of LAV-1_{BRU}

[b]Underlined residues indicate residues conserved between HIV-1 and HIV-2. Dashes indicate sequence identity with LAV-1_{BRU}

[c]The three-letter and single-letter abbreviations for the amino acids are: Ala (A), alanine; Arg (R), arginine; Asn (N), asparagine; Asp (D), aspartic acid; Cys (C), cysteine; Gln (Q) glutamine; Glu (E), glutamic acid; Gly (G), glycine; His (H), histidine; Ile (I), isoleucine; Leu, leucine; Lys (K), lysine; Met (M), methionine; Phe (F), phenylalanine; Pro (P), proline; Ser (S), serine; Thr (T), threonine; Trp (W), tryptophan; Tyr (Y), tyrosine; Val (V), valine

Table 3. Characterization of selected sera from Zairian AIDS patients. [Adapted with permission of Gnann et al. (1987c); copyright 1987 by the AAAS]

Specimen	HIV culture	Western blot (HIV-1)	Enzyme-linked immunosorbent assays					
			Whole virus HIV-1	LGLWGCSGKLIC HIV-1 (Zaire)	LGIWGCSGKLIC HIV-1 (France, US)	LGIWGCSGKHIC HIV-1 (Zaire)	LGMWGCSGKHIC HIV-1 (Zaire)	NSWGCAFRQVC HIV-2 (CVI)
Z-9	+	p24 +, p41 + / −	+	−[a]	+	+	+	−
Z-23	+	p24 +, p41 + / −	+	−	−	+	+	−
Z-24	+	p25 +	+	+	+	+	+	−
Z-29	+	ND[b]	+	+	+	−	−	−
Z-31	+	p24 +, p41 +	+	+	+	+	+	−
Z-34	+	p24 +, p41 −	+	−	−	+	+	−
Z-47	+	p24 +, p41 −	+	−	−	+	+	−
Z-50	−	p24 +, p41 +	+	−	−	+	+	ND
Z-52	+	p25 +, p41 +, p60 +	+	+	+	+	+	−
Z-68	+	ND	+	+	+	−	+	−
Z-72	+	p25 +	+	+	+	+	+	−
Z-75	+	p18 +, p25 +, p41 +, p60 +	+	+	+	+	+	−
Z-81	+	p18 +, p25 +, p41 +, p60 +	+	+	+	+	+	−

[a] +, positive; −, negative (the cutoff for positivity for the peptide ELISA was defined as the mean OD_{492} plus 3 standard deviations for a panel of 24 negative control sera. Specimens were scored as positive or negative at a serum dilution of 1:64)

[b] ND, not done, insufficient quantity of specimen

been sequenced (Table 2). However, some strains with substitutions in this region at amino acid positions 600 and 607 have been described, including isolates from Zaire. In particular, nonconservative substitutions were noted in Zairian isolates LAV-1_{MAL} and LAV-1_{ELI} (Table 2). We speculated that antibody from an individual infected with one of these variant virus strains might not recognize the original peptide but should bind specifically with peptides containing the appropriate amino acid substitutions. We tested this hypothesis by synthesizing additional peptides corresponding to the sequences of isolates LAV-1_{MAL} and LAV-1_{ELI}, as shown in Table 2 (GNANN et al. 1987c). The five sera from Zairian AIDS patients that failed to react with the original peptide reacted strongly with the peptides corresponding to the Zairian HIV-1 isolates (Table 3). Conversely, two sera that reacted with the original peptide failed to react with the substituted peptides, whereas the remainder of the sera reacted with all three peptides. These data indicate that synthetic peptide antigens can be constructed to detect strain-specific antibody responses and that immune recognition can be altered by substitution of a single amino acid in a critical position.

6 Antigenic Epitope on the Transmembrane Glycoprotein of HIV-2

While HIV-1 and HIV-2 are related retroviruses, they are quite distinct at both genetic and antigenic levels (GUYADER et al. 1987). We reasoned, however, that the region of the transmembrane glycoprotein shown to be immunodominant for HIV-1 might also be immunologically important for HIV-2. When the sequence of the two glycoproteins were compared, identical amino acids were noted at 5 of the 12 positions (Table 2), including the 2 cysteine residues previously shown to be essential for antigenicity of the HIV-1 epitope. An 11-residue peptide (NSWGCAFRQVC) corresponding to amino acids 591–603 of HIV-2_{ROD} was synthesized and found to be highly reactive with sera from HIV-2-infected patients but not with sera from HIV-seronegative individual or with sera from patients infected with HIV-1 only (GNANN et al. 1987c). Other investigators have reported a 22-amino-acid peptide derived from the corresponding region of simian immunodeficiency virus (a retrovirus closely related to HIV-2) that also reacts with sera from HIV-2-infected individuals (NORRBY et al. 1987).

We have subsequently used the HIV-1 and HIV-2 peptides to test sera from over 200 AIDS patients living in areas of West Africa where both viruses are endemic (J. GNANN, K. DECOCK, J. MCCORMICK, M. OLDSTONE 1988, unpublished observations). The peptide enzyme immunoassays permit most sera to be readily classified as positive for HIV-1 or HIV-2, and the results show excellent correlation with those from Western blotting. However, there is a subgroup of West African sera that show reactivity with both HIV-1 and HIV-2 antigens by peptide ELISA and immunoblotting. We have not noted this phenomenon of dual reactivity when testing AIDS sera from the United States or Central Africa. Possible explanations include immunologic cross-reactivity, dual infection, or infection with an unidentified intermediate virus type. These dually reactive sera remain the subject of active investigation.

7 Future Directions

Synthetic peptide technology has been used to identify several antigenic epitopes on the envelope glycoproteins of HIV-1. An immunodominant site on the transmembrane glycoprotein can be represented by a synthetic peptide with the sequence LGLWGCSGKLIC. The peptide can be used as an antigen in a sensitive and specific enzyme immunoassay to identify antibodies in sera from HIV-1-infected patients. The minimal sequence for antibody recognition is a 7-amino-acid sequence containing two essential cysteine residues that apparently confer proper antigenic conformation by formation of a disulfide bond.

Evaluation of HIV-1-positive sera from African patients demonstrated a method for serological distinguishing infection caused by different HIV-1 strains. An immunodominant epitope was identified that was largely conserved among strains, yet exhibited residue variability at certain key positions. The design and synthesis of peptides containing these amino acid substitutions allowed detection of very specific antibody responses. If additional viral epitopes with these characteristics (immunodominant, generally conserved, with limited strain-specific substitutions) are uncovered, a panel of synthetic peptides could be designed for serologic classification of the infecting HIV-1 strains.

Investigations with the HIV-2 peptide illustrate a different principle. Although the corresponding glycoprotein regions represented by the HIV-1 and HIV-2 peptides share only limited amino acid sequence homology, they are both strongly and specifically antigenic. The two regions probably have similar three-dimensional conformations and functional roles in the respective viruses (although the function remains undefined). If the HIV types are identified, they are likely to share this characteristic of an antigenic domain near the amino terminus of the transmembrane glycoprotein that contains two cysteine residues and is presented as a cyclic epitope (although the overall amino acid sequence may be very different). In addition to preparing synthetic peptides for distinguishing HIV-1 and HIV-2 infections, it may be possible to custom-design reagents that combine antigenic features of both viruses and, thereby, diagnose infection caused by either HIV-1 or HIV-2.

Acknowledgments. This is Publication Number 5431-IMM from the Department of Immunology, Scripps Clinic and Research Foundation, La Jolla, CA 92037. This work was supported in part by U.S. Public Health Service grants AI-07007, NS-12428, training grant T32-NS-07078 (JWG) from the National Institutes of

Health and the U.S. Army Medical Research and Development Command under Contract No. DAMD17-88-C-8103. Opinions, interpretations, conclusions, and recommendations are those of the author and are not necessarily endorsed by the U.S. Army.

References

Alizon M, Wain-Hobson S, Montagnier L, Sonigo P (1986) Genetic variability of the AIDS virus: nucleotide sequence analysis of two isolates from African patients. Cell 46: 63–74

Benn S, Rultedge R, Folks T, Gold J, Baker L, McCormick J, Feorino P, Piot P, Quinn T, Martin MA (1985) Genomic heterogeneity of AIDS retroviral isolates from North America and Zaire. Science 230: 949–951

Brun-Vezinet F, Rouzioux C, Montagnier L, Chamaret S, Gruest J, Barre-Sinoussi F, Gerolid D, Chermann JC, McCormick J, Mitchell S, Piot P, Taelman H, Mirlangu KB, Wobin O, Mbendi N, Mazebo P, Kalambayi K, Bridts C, Desmyter J, Feinsod FM, Quinn TC (1984) Prevalence of antibodies to lymphadenopathy associated retrovirus in African patients with AIDS. Science 226: 453–456

Brun-Vezinet F, Rey MA, Katlama C, Girard PM, Roulot D, Yeni P, Lenoble L, Clavel F, Alizon M, Gadelle S, Madjar JJ, Harzic M (1987) Lymphadenopathy associated virus type 2 in AIDS and AIDS-rated complex: clinical and virological features in four patients. Lancet i: 128–132

Cheng-Mayer C, Rutka JT, Rosenblum ML, McHugh T, Stites DP, Levy JA (1987) Human immunodeficiency virus can productively infect cultured human glial cells. Proc Natl Acad Sci USA 84: 3526–3530

Cheng-Mayer C, Seto D, Tateno M, Levy JA (1988) Biological features of HIV-1 that correlate with virulence in the host. Science 240: 80–82

Chou PY, Fasman GD (1978) Prediction of the secondary structure of proteins from their amino acid sequence. Adv Enzymol 47: 45–148

Clavel F, Guetard D, Brun-Vezinet F, Chamaret S, Rey MA, Santos-Ferreira MO, Laurent AG, Dauguet C, Katlama C, Rouzioux C, Klatzman D, Champalimaud JL, Montagnier L (1986) Isolation of a new human retrovirus from West African patients with AIDS. Science 233: 343–346

Clavel F, Mansinho K, Chamaret S, Guetard D, Favier V, Nina J, Santos-Ferreira MO, Champalimaund JL, Montagnier L (1987) Human immunodeficiency virus type 2 infection associated with AIDS in West Africa. N Engl J Med 316: 1180–1185

Coffin JM (1986) Genetic variation in AIDS viruses. Cell 46: 1–4

Dahl K, Martin K, Miller G (1987) Differences among human immunodeficiency virus strains in their capacities to induce cytolysis or persistent infection of a lymphoblastoid cell line immortalized by Epstein-Barr virus. J Virol 61: 1602–1608

Desai SM, Kalyanaraman VS, Casey JM, Srinivasan A, Andersen PR, Devare SG (1986) Molecular cloning and primary nucleotide sequence analysis of a distinct human immunodeficiency virus isolate reveal significant divergence in its genomic sequences. Proc Natl Acad Sci (USA) 83: 8380–8384

Evans LA, McHugh TM, Stites DP, Levy JA (1987) Differential ability of human immunodeficiency virus isolates to productively infect human cells. J Immunol 138: 3415–3418

Gnann JW, Schwimmbeck PL, Nelson JA, Truax AB, Oldstone MBA (1987a) Diagnosis of AIDS using a 12 amino acid peptide representing an immunodominat epitope of human immunodeficiency virus. J Infect Dis 156: 261–267

Gnann JW, Nelson JA. Oldstone MBA (1987b) Fine mapping of an immunodominant domain in the transmembrane glycoprotein of human immunodeficiency virus. J Virol 61: 2639–2641

Gnann JW, McCormick JB, Mitchell S, Nelson JA, Oldstone MBA (1978c) Synthetic peptide immunoassay distinguishes HIV type 1 and type 2 infections. Science 237: 1346–1349

Gnann JW, Smith LL, Oldstone MBA (1989) Custom-designed synthetic peptide immunoassays for distinguishing HIV type 1 and type 2 infections. In: Langone JL (ed) Methods in enzymology: immunochemical techniques—anti-idiotypic antibodies and molecular mimicry, Academic Press, Orlando, FL, pp. 693–714.

Gurgo C, Guo H-G, Franchini G, Aldovini A, Collalti E, Farrell K, Wong-Staal F, Gallo RC, Reitz MS (1988) Envelope sequences of two new United States HIV-1 isolates. Virology 164: 531–536
Guyader M, Emerman M, Sonigo P, Clavel F, Montagnier L, Alizon M (1987) Genome organization and transactivation of the human immunodeficiency virus type 2. Nature 326: 622–669
Hahn BH, Gonda MA, Shaw GM, Popovic M, Hoxie J, Gallo RC, Wong-Staal F (1985) Genomic diversity of the AIDS virus HTLV-III: different viruses exhibit greatest divergence in their envelope genes. Proc Natl Acad Sci USA 82: 4813–4817
Hopp TP, Woods KR (1981) Prediction of protein antigenic determinants from amino acid sequences. Proc Natl Acad Sci USA 78: 3824–3828
Kennedy RC, Henkel RD, Pauletti D, Allan JS, Lee TH, Essex M, Dreesman GR (1986) Antiserum to a synthetic peptide recognizes the HTLV-III envelope glycoprotein. Science 231: 1556–1559
Lerner RA (1984) Antibodies of predetermined specificity in biology and medicine. Adv Immunol 36: 1–44
Modrow S, Hahn BH, Shaw GM, Gallo RC, Wong-Staal F, Wolf H (1987) Computer assisted analysis of envelope protein sequences of seven human immunodeficiency virus isolates: prediction of antigenic epitopes in conserved and variable regions. J Virol 61: 570–578
Norrby E, Biberfeld G, Chiodi F, von Gegerfeldt A, Naucler A, Parks E, Lerner R (1987) Discrimination between antibodies to HIV and to related retroviruses using site-directed serology. Nature 329: 248–250
Palker TJ, Matthews TJ, Clark ME, Cianciolo GJ, Randall RR, Langlois AJ, White GC, Safai B, Snyderman R, Bolognesi DP, Haynes BF (1987) A conserved region at the COOH terminus of human immunodeficiency virus gpl20 envelope protein contains an immunodominant epitope. Proc Natl Acad Sci USA 84: 2479–2483
Palker TJ, Clark ME, Langlois AJ, Matthews TJ, Weinhold KJ, Randall RR, Bolognesi DP, Haynes BF (1988) Type-specific neutralization of the human immunodeficiency virus with antibodies to *env*-encoded synthetic peptides. Proc Natl Acad Sci USA 85: 1932–1936
Pauletti D, Simmonds R, Dreesman GR, Kennedy RC (1985) Application of a modified computer algorithm in determining potential antigenic determinants associated with the AIDS virus glycoprotein. Anal Biochem 151: 540–546
Prince AM, Pascual D, Kosolapov LB, Kurokawa D, Baker L, Rubinstein P (1987) Prevalence, clinical significance, and strain specificity of neutralizing antibody to the human immunodeficiency virus. J Infect Dis 156: 268–272
Ratner L, Haseltine W, Patarca R, Livak KJ, Starich B, Josephs SF, Doran FR, Rafalski JA, Whitehorn EA, Baumeister K, Ivanoff L, Petteway SR, Pearson ML, Lautenberger JA, Papas TS, Ghrayeb J, Chang NT, Gallo RC, Wong-Staal F (1985) Complete nucleotide sequence of the AIDS virus, HTLV-III. Nature 313: 277–284
Sanchez-Pescador R, Power MD, Barr PJ, Steimer KS, Stempien MM, Brown-Shimer SL, Gee WW, Renard A, Randolph A, Levy JA, Dina D, Luciw PA (1985) Nucleotide sequence and expression of an AIDS-associated retrovirus (ARV-2). Science 227: 484–492
Schrier RD, Gnann JW, Langlois AJ, Shriver K, Nelson JA, Oldstone MBA (1988) B and T lymphocyte responses to an immunodominant epitope of human immunodeficiency virus. J Virol 62: 2531–2536
Smith RS, Naso RB, Rosen J, Whalley Am Hom YL, Hoey K, Kennedy CJ, McCutchan JA, Spector SA, Richman DD (1987) Antibody to a synthetic oligopeptide in subjects at risk for human immunodeficiency virus infection. J Clin Microbiol 25: 1498–1504
Srinivasan A, Anand R, York D, Ranganathan P, Feorino P, Schochetman G, Curran J, Kalyanaraman VS, Luciw PA, Sanchez-Pescador R (1987) Molecular characterization of human immunodeficiency virus from Zaire: nucleotide sequence analysis identifies conserved and variable domains in the envelope gene. Gene 52: 71–82
Sternberg MJ, Barton GJ, Zvelebil MJ, Cookson J, Coates ARM (1987) Prediction of antigenic determinants and secondary structures of the major AIDS virus proteins. FEBS Lett 218: 231–237
Starcich BR, Hahn BH, Shaw GM, McNeely PD, Modrow S, Wolf H, Parks ES, Parks WP, Josephs SF, Gallo RC, Wong-Staal F (1986) Identification and characterization of conserved and variable regions in the envelope gene of HTLV-III/LAV, the retrovirus of AIDS. Cell 45: 637–748
Sutcliffe JG, Shinnick TM, Green N, Lerner RA (1983) Antibodies that react with predetermined sites on proteins. Science 219: 660–666
Tainer JA, Getzoff ED, Alexander H, Houghten RA, Olson AJ, Lerner RA, Hendrickson WA (1984) The reactivity of antipeptide antibodies is a function of the atomic mobility of sites in a protein. Nature 312: 127–134

Wain-Hobson S, Sonigo P, Danos O, Cole S, Alizon M (1985) Nucleotide sequence of the AIDS virus, LAV. Cell 40: 9–17
Walter G, Scheidtmann KH, Carbone A, Laudana AP, Doolittle RF (1980) Antibodies specific for the carboxy- and amino-terminal regions of simian virus 40 large antigen. Proc Natl Acad Sci USA 77: 5197–5200
Wang JJG, Steel S, Wisniewolski R, Wang CY (1986) Detection of antibodies to human T-lymphotropic virus type III by using a synthetic peptide of 21 amino acid residues corresponding to a highly antigenic segment of gp41 envelope protein. Proc Natl Acad Sci USA 83: 6159–6163
Westhof E, Altschuh D, Moras D, Bloomer AC, Mondragon A, Klug A, Van Regenmortel MHV (1984) Correlation between segmental mobility and the location of antigenic determinants in proteins. Nature 311: 123–126
Willey RL, Rutledge RA, Dias S, Folks T, Theodore T, Buckler CE, Martin MA (1986) Identification of conserved and divergent domains within the envelope gene of the acquired immunodeficiency syndrome retrovirus. Proc Natl Acad Sci (USA) 83: 5038–5042

Human Immunodeficiency Virus Infection of the CNS: Characterization of "Neurotropic" Strains

C. CHENG-MAYER and J. A. LEVY

1 Introduction

Neurologic abnormalities have been reported in over 50% of patients with the acquired immune deficiency syndrome (AIDS), and in some cases can be the only clinical manifestation of infection with the human immunodeficiency virus (HIV) (SNIDER et al. 1983; NIELSON et al. 1984; PETITO et al. 1985; LEVY et al. 1985; NAVIA et al. 1986a, b; DE LA MONTE et al. 1987, GABUZDA and HIRSCH 1987; NAVIA and PRICE 1987; ELDER and SEVER 1988; PRICE et al. 1988; ROSENBLUM et al. 1988). Substantial evidence suggests that many of the neurologic symptoms, particularly vacuolar myelopathy and progressive dementia, are the direct consequence of HIV infection of the CNS (EPSTEIN et al. 1984–85; HO et al. 1985a; LIPKIN et al. 1985; ANDERS et al. 1986; SHARER et al. 1986; CORNBLATH et al. 1987). This apparent capacity of HIV to induce neuropathology is not surprising, since other members of the lentivirus family (for example, visna and caprine arthritis encephalitis virus) are also "neurotropic". In the case of visna, the virus has been shown to infect cells of the choroid plexus, alveolar macrophages/monocytes, astrocytes, and oligodendrocytes, both in infected animals and in tissue culture (BRAHIC et al. 1984; NARAYAN et al. 1985; GENDELMAN et al. 1986; HASSE 1986). However, the neurotropism of HIV raises

Cancer Research Institute, University of California, School of Medicine, San Francisco, CA 94143, USA

Current Topics in Microbiology and Immunology, Vol. 160

important questions as to (a) whether specific "neurotropic" strains of HIV exist, (b) what cells in the brain are infected, and (c) what genetic factors of "neurotropic" HIV impart the ability to infect cells of the CNS.

2 Central Nervous System HIV Infection

2.1 Pathology Studies

The presence of HIV in the brain has been convincingly demonstrated in a series of studies using different molecular, immunohistochemical, virologic, and immunologic techniques. Furthermore, HIV has been isolated readily from brain biopsy material and CSF of patients with neurologic symptoms. It has also been recovered from the CSF of asymptomatic individuals (see below). By Southern blot analysis, HIV-specific DNA was shown to be present in brain tissues of AIDS patients with dementia in quantities comparable to that present in lymphoid tissues (Shaw et al. 1985; Harper et al. 1986). In situ hybridization, immunohistochemical studies, and electron microscopic analysis of brain sections have revealed the production of viral RNA, antigens, and particles in the brain (Gabuzda et al. 1986; Koenig et al. 1986; Stoler et al. 1986; Wiley et al. 1986; Gyorkey et al. 1987; Meyenhofer et al. 1987; Pumarola-Sune et al. 1987) (Table 1). Microglia- and macrophages are the primary cell types in the brain

Table 1. Detection of HIV in brain tissues

Method of detection	Cell types infected	Reference
In situ	Macrophages, multinucleated giant cells	Koenig et al. (1986) Sharer et al. (1986) Stoler et al. (1986) Wiley et al. (1986)
	Endothelial cells, oligodendroglial	Wiley et al. (1986) Levy et al. (1987)
Immunohistochemical	Macrophages, multinucleated giant cells	Gabuzda et al. (1986) Koenig et al. (1986) Wiley et al. (1986) Pumarola-Sune et al. (1987)
	Endothelial cells, oligodendroglial, astroglial, neurons	Wiley et al. (1987) Gyorkey et al. (1987) Wiley et al. (1986) Pumarola-Sune et al. (1987)
Electron microscopy	Macrophages, multinucleated giant cells	Epstein et al. (1984–85) Gyorkey et al. (1985) Koenig et al. (1986) Meyenhofer et al. (1987)
	Oligodendroglial, astroglial	Gyorkey et al. (1987) Epstein et al. (1984–85)

reported to be infected (GYORKEY et al. 1985; DICKSON 1986; KOENIG et al. 1986; SHARER et al. 1986; VAZEUX et al. 1987), although endothelial cells, and, much less frequently, astrocytes, oligodendrocytes, and neurons, have also been shown to be infected (EPSTEIN et al. 1984–85; WILEY et al. 1986; GYORKEY et al. 1987; LEVY et al. 1987).

2.2 Recovery of HIV from CNS

Infectious viruses have been isolated from brain tissues and CSF of HIV-infected individuals (HO et al. 1985b; LEVY et al. 1985; CHIODI et al. 1986; GARTNER et al. 1986a; HOLLANDER and LEVY 1987; SONNERBORG et al. 1987) (Table 2). The isolation of HIV, together with immune detection of antibodies to HIV and the presence of "oligoclonal" immunoglobulin bands in CSF, may constitute early signs of CNS infection (RESNICK et al. 1985; GOUDSMIT et al. 1986; EPSTEIN et al. 1987). In our laboratory, over 40 HIV isolates have been recovered from the

Table 2. Recovery of HIV from CSF and brain tissues of patients with and without neurologic symptoms

Neurologic complication	No. of patients	No. with HIV
A. LEVY et al. (1985)		
Neurologic symptoms	12	12
None	2	1
Total:	14	13
B. HO et al. (1985b)		
Meningitis	9	7
Dementia	16	10
Myelopathy	4	4
Peripheral neuropathy	3	3
Others (e.g., lymphoma, toxoplasmosis)	6	5
Unrelated	3	0
None	9	0
Total:	50	29
C. CHIODI et al. (1986)		
None	1	1
D. GARTNER et al. (1986a)		
Dementia	1	1
E. HOLLANDER and LEVY (1987)		
Meningitis	3	2
Subacute encephalopathy	9	6
Myelopathy	1	1
Peripheral neuropathy	4	2
Others (tumors, secondary infection)	12	9
None	8	5
Total:	37	25
F. SONNERBORG et al. (1987)		
Neurologic symptoms	23	15
None	48	29
Total:	71	44

choroid plexus, cerebellum, spinal cord, cerebral cortex, and CSF of infected individuals with and without neurologic symptoms. In some cases, viruses have been recovered from CSF and blood of the same individual; in others, HIV has been found in only one of these compartments (HOLLANDER and LEVY 1987). The various viruses are being analyzed to determine whether a neurotropic subtype of HIV exists (Sect. 3).

2.3 Infection of Brain Cells In Vitro

As noted above, the major cell found infected by HIV in the CNS is the brain macrophage (microglia). In tissue culture, HIV has been shown productively to infect primary monocyte/macrophages (GARTNER et al. 1986b; HO et al. 1986; NICHOLSON et al. 1986; SALAHUDDIN et al. 1986; CROWE et al. 1987; KOYANAGI et al. 1987; CHENG-MAYER et al. 1988b, c). However, the characteristics of HIV replication in monocyte/macrophages differ from that in T cells. Infection of T cells often results in cytopathic effects (CPEs) which are manifested as syncytia or multinucleated-giant-cell formation concomitant with release of high titers of infectious virus. Induction of CPEs is dependent on the interaction of CD4 (the major viral receptor on the cell) and the viral envelope glycoprotein gp120 (HOXIE et al. 1986; LIFSON et al. 1986a, b, c; SODROSKI et al. 1986) and leads to the downregulation or disappearance of the CD4 molecule on the infected cell surface. In contrast, CPEs and release of high titer virus are only observed in infection of primary monocytes/macrophages by a limited number of HIV strains (CROWE et al. 1987; CHENG-MAYER et al. 1988b, c; WEISS and LEVY, unpublished observation). In general, only transient CPEs are observed. Furthermore, often very few viral particles are released or found associated with the plasma membrane; the majority of viral particles accumulate in intracytoplasmic vacuoles (MEYENHOFER et al. 1987; GENDELMAN et al. 1988a; ORENSTEIN et al. 1988). These findings may explain the lack of substantial cytopathology seen in infected monocytes/macrophages and suggest that this cell population could serve as a primary reservoir for virus, thereby enhancing persistence and spread within the infected host.

Similarly, infection of glial cell cultures in vitro is noncytopathic with release of only low levels of infectious virions for protracted periods (CHENG-MAYER et al. 1987; CHIODI et al. 1987; CHRISTOFINIS et al. 1987; DEWHURST et al. 1987a, b; KOYANAGI et al. 1987; DEWHURST et al. 1988). The CD4 molecule appears to be necessary for infection of T cells, monocyte/macrophages, and some glial cells (DALGLEISH et al. 1984; KLATZMAN et al. 1984; MADDEN et al. 1986; MCDOUGAL et al. 1986; NICHOLSON et al. 1986; CROWE et al. 1987; DEWHURST et al. 1987a). However, there are glial cell lines which are susceptible to HIV infection but do not express surface CD4 or mRNA for this receptor (CHENG-MAYER et al. 1987). Other glial cell lines express CD4 mRNA, but addition of antibody to CD4 or soluble CD4 does not block infection (HAROUSE et al. 1988; WEBER et al. 1988).

These results suggest that other viral entry mechanisms may be involved in HIV infection of cells in the CNS.

3 Unique Properties of CNS-Derived HIV

Heterogeneity among HIV isolates is now a well-recognized phenomenon. Differences have been observed in genomic structure of HIV isolates as well as in their biologic and serologic properties (BENN et al. 1985; WONG-STAAL et al. 1985; HAHN et al. 1985, 1986; COFFIN 1986; ASJO et al. 1987; LEVY et al. 1987; CHENG-MAYER et al. 1988a, b, c; LEVY 1988, SAAG et al. 1988; TATENO and LEVY 1988; TERSMETTE et al. 1988). In our laboratory, we have been able to distinguish individual HIV by differences in cell tropism, kinetics of replication, cytopathology, and susceptibility to serum neutralization. Whereas all HIV isolates will grow in peripheral mononuclear cells (PMCs), differences were observed in their ability to productively infect human T-cell lines (HUT 78, CEM, Jurkat), monocytic cell lines, transformed B-cell lines, and primary monocyte/macrophages (EVANS et al. 1987; LEVY et al. 1987; CHENG-MAYER et al. 1988b, c). They also differed in their kinetics of replication in the infected cells. Some grew rapidly and to high titers, inducing a high degree of cytopathology with reduced expression of the CD4 receptor molecule. Others grew slowly and were less cytopathic (LEVY et al. 1987; CHENG-MAYER et al. 1988b). Similar observations have been made by ASJO et al. (1986). The ability of an HIV isolate to grow rapidly and to induce cytopathology was recently found to correlate with their ability to form plaques in the MT-4 cell line (TATENO and LEVY 1988). Thus the plaque assay, first described by HARADA and his coworkers (1985) for HIV, can be used to identify "cytopathic" or "virulent" strains of HIV. In addition, serologic subtypes of HIV have been identified which display different degrees of sensitivity to serum neutralization (CHENG-MAYER et al. 1988a).

These patterns of differential biologic and serologic properties were used to determine whether isolates from the brain might constitute a distinct subgroup of HIV. In our studies of seven isolates from brain and eight isolates from peripheral blood, brain isolates were found to be less cytopathic and display a greater tropism for primary monocytes/macrophages (CHENG-MAYER and LEVY 1988; CHENG-MAYER et al. 1988c, d) (Table 3). Putative "neurotropic" variants with distinct cytopathology and cell tropism properties have also been reported by others (GARTNER et al. 1986b; KOYANAGI et al. 1987; ANAND 1988; ANAND et al. 1987). In addition, we have shown that, in contrast to peripheral blood isolates, the brain isolates do not productively infect established human T-cell lines, do not substantially reduce expression of the CD4 receptor molecule in infected CD4-positive cells, and are less susceptible to serum neutralization (CHENG-MAYER and LEVY 1988; CHENG-MAYER et al. 1988c, 1989). These data suggest

Table 3. Comparison of blood and brain isolates

Properties	Blood	Brain	Reference
Replication in T-cell lines	+ +	−	Cheng-Mayer and Levy (1988), Cheng-Mayer et al. (1988c; 1989)
Replication in primary monocyte/macrophages	+	+ +	Gartner et al. (1986b) Cheng-Mayer et al. (1988c; 1989)
Cytopathology in $T4^+$ cells	+ +	−	Anand (1988) Cheng-Mayer et al. (1988c, d)
Reduced CD4 expression	+ +	±	Cheng-Mayer et al. (1988d)
Plaque formation in MT4 cells	+	−	Cheng-Mayer et al. (1988c, d), Tateno and Levy (1988)
Serum neutralization	+ +	±	Cheng-Mayer and Levy (1988), Cheng-Mayer et al. (1988c; 1989)

For replication, + + indicates efficient and high titers of replication by > 60% isolates tested; + indicates efficient replication by < 50% isolates tested; − indicates absence of replication. For cytopathology, + + indicates presence of ballooning and syncytia formation in infected cells; − absence of cytopathic effects. For reduction in CD4 antigen expression, + + indicates > 80% reduction, ± indicates < 20% reduction. For serum neutralization, + + indicates susceptibility to neutralization by three high-titered HIV-positive sera; ± indicates neutralization by one out of the three sera tested

that brain isolates constitute a distinct serologic subgroup of HIV with shared biologic characteristics that are different from non-CNS isolates.

When paired CSF and blood isolates from the same individuals were examined, they were found to be genomic variants of each other as determined by restriction endonuclease mapping (Cheng-Mayer et al. 1988c, 1989d). When their biologic properties were compared, minor differences were noted in their ability to infect primary monocytes/macrophages and a variety of glial cell lines bearing glial fibrillary acid protein, a marker specific for astrocytes (Rutka et al. 1987) (Table 4). Both the CSF and blood isolates established low-level persistent infection of non-T cells, but the kinetics of replication and titers of viruses produced were different. The brain isolates replicated better in primary monocyte/macrophages whereas the blood isolates infected glial cells more

Table 4. Comparison of blood and brain isolates from the same individual. [Table in part from Cheng-Mayer et al. (1988c)]

Isolate (HIV_{SF})	Replication in T-Cell lines	mϕ	glial	Cytopathology	Reduced CD4 expression	Plaque formation
128A (spinal cord)	−	+	±	−	−	−
128B (lymph node)	−	±	+	−	−	−
185A (CSF)	−	+	±	−	−	−
185B (PBL)	−	±	+	−	−	−

CSF, cerebrospinal fluid; PBL, peripheral blood lymphocytes or mononuclear cells; mϕ, macrophage. For replication in macrophages, + indicates RT activity > 100000 cpm/ml; ± indicates RT activity < 50,000 cpm/ml. For replication in glial cells, + indicates RT activity > 100000 cpm/ml in PBL cocultivated with infected glial cells and ± indicates RT activity < 20,000 cpm/ml when measured in the mixed culture efficiently. Genotypically distinct viral isolates obtained from the same individual with differences in cellular tropism has been reported by others (KOYANAGI et al. 1987). In this latter study, the isolate from the brain efficiently infected primary monocyte/macrophages whereas virus from the CSF productively infected a glioma cell line.

4 Viral Pathogenesis

The finding that macrophages are the major cell type in the brain to be infected has led to the suggestion these cells play a central role in CNS pathogenesis. The infected macrophages may act as vehicles of transport of virus from the peripheral circulation across the blood-brain barrier, and as reservoirs of infection in the brain parenchyma. Dissemination of virus infection to neighboring neurons or glia might somehow kill or interfere with the normal function of these cells (LEE et al. 1987). Alternatively, infected macrophages might secrete soluble factor(s) that cause glial or neuronal cell dysfunction (GENDELMAN et al. 1988b). However, a low-level, persistent infection of these glial or neuronal cells by HIV leading directly or indirectly to neuropathology should also be considered (LEVY 1988).

The unique properties of the brain isolates described above (Sect. 3) might explain their ability to induce disease in the CNS. It would be desirable to identify which viral genes govern the "neurotropism" of HIV, as defined by their ability to infect macrophages and other cells of the CNS. This question is being approached by constructing recombinants between molecularly cloned blood isolates (HIV_{SF2}, HIV_{SF33}) and brain isolates (HIV_{SF128A}, HIV_{SF162}). The two blood isolates differ in the degree of cytopathology induced in infected T cells, in plaque formation, and in kinetics of replication (TATENO and LEVY 1988). Genomic recombinants have already been generated between HIV_{SF2} and HIV_{SF128A} and between HIV_{SF2} and HIV_{SF33}. The viruses produced are being tested for their biologic and serologic properties. Preliminary data with recombinant viruses in the envelope region suggest that this gene plays a role in controlling cell tropism and cytopathology in vitro (CHENG-MAYER et al., unpublished observations). Similar findings have been reported recently (FISHER et al. 1988) using genotypically related viruses. Recombinants generated in other region(s) of the virus should reveal additional genes that control kinetics and levels of replication and its neurotropism.

5 Conclusions

Similar to other members of the lentivirus family, HIV has been shown to infect and cause disease in the CNS. The predominant cell type observed to be infected in the brain is the macrophage and derivative multinucleated giant cells.

However, low levels of infection are also detected in glial cells, endothelial cells, and perhaps neurons. Putative "neurotropic" HIV variants have been isolated that share properties, defined by cell tropism, cytopathology, and antigenicity, that distinguish them from blood-derived HIV strains. Studies are in progress to dissect the viral determinants that control the unique characteristics of these putative "neurotropic" variants. However, in the absence of an animal model test system, the importance of these viral genetic factors in determining infection of various cell types in the brain and causing CNS pathology cannot yet be directly addressed.

Acknowledgments. The research by the authors cited in this paper was supported by grants from the National Institutes of Health R29-NS25007, P01-A124286, and R01-A124499.

References

Anand R (1988) Natural variants of human immunodeficiency virus from patients with neurological disorders do not kill T4$^+$ cells. Ann Neurol 23: s66–s70

Anand R, Siegal F, Reed C, Cheung T, Forlenza S, Moore J (1987) Noncytocidal natural variants of human immunodeficiency virus isolated from AIDS patients with neurological disorders. Lancet ii: 234–238

Anders KH, Guerra WF, Tomiyasu U, Verity MA, Vinters HV (1986) The neuropathology of AIDS: UCLA experience and review. Am J Pathol 124: 537–558

Asjo B, Morfeldt-Manson L, Albert J, Biberfeld G, Karlson A, Lidman K, Fenyo EM (1986) Replicative capacity of human immunodeficiency virus from patients with varying severity of HIV infection. Lancet II: 660–662

Benn S, Rutledge R, Folks T, Gold J, Baker L, McCormick J, Feorino P, Piot P, Quinn T, Martin MA (1985) Genomic heterogeneity of AIDS retroviral isolates from North America and Zaire. Science 230: 949–951

Brahic M, Hasse AT, Cash E (1984) Simultaneous *in situ* detection of viral RNA and antigens. Proc Natl Acad Sci USA 81: 5445–5448

Cheng-Mayer C, Levy JA (1988) Distinct biological and serological properties of human immunodeficiency viruses from the brain. Ann Neurol 23: s58–s61

Cheng-Mayer C, Rutka J, Rosenblum M, McHugh C, Stites D, Levy JA (1987) Human immunodeficiency virus can productively infect cultured human glial cells. Proc Natl Acad Sci USA 84: 3526–3530

Cheng-Mayer C, Homsy J, Evans LA, Levy JA (1988a) Identification of human immunodeficiency virus subtypes with distinct patterns of sensitivity to serum neutralization. Proc Natl Acad Sci USA 85: 2815–2819

Cheng-Mayer C, Seto D, Tateno M, Levy JA (1988b) Biologic features of HIV-1 that correlate with virulence in the host. Science 240: 80–82

Cheng-Mayer C, Tateno M, Seto D, Levy JA (1988c) Distinguishing features of neurotropic human immunodeficiency viruses. Vaccine 88: 259–264

Cheng-Mayer C, Weiss C, Seto D, Levy JA (1989) Isolates of human immunodeficiency virus type 1 from the hair may constitute a special group of the AIDS virus. Proc Natl Acad Sci USA 86: 8575–8579

Chiodi F, Asjo B, Fenyo EM, Norkrans G, Hagberg L, Albert J (1986) Isolation of human immunodeficiency virus from cerebrospinal fluid of antibody-positive virus carrier without neurological symptoms. Lancet ii: 1276–1277

Chiodi F, Fuerstenberg S, Gidlund M, Asjo B, Fenyo EM (1987) Infection of brain-derived cells with the human immunodeficiency virus. J Virol 61: 1244–1247

Christofinis G, Papadaki L, Sattentau Q, Ferns RB, Tedder R (1987) HIV replicates in cultured human brain cells. AIDS 1: 229–234
Coffin JM (1986) Genetic variation in AIDS viruses. Cell 46: 1–4
Cornblath DR, McArthur JC, Kennedy PGE, Witte AS, Griffin JW (1987) Inflammatory demyelinating peripheral neuropathies associated with human T-cell lymphotropic virus type III infection. Ann Neurol 21: 32–40
Crowe S, Mills J, McGrath MS (1987) Quantitative immunocytofluorographic analysis of CD4 surface antigen expression and HIV infection of human peripheral blood monocyte/macrophages. AIDS Res Hum Retrovirus 3: 135–145
Dalgleish AG, Beverley PCL, Clapham PR, Crawford DH, Greaves MF, Weiss RA (1984) The CD4 (T4) antigen is an essential component of the receptor for the AIDS retrovirus. Nature 312: 763–767
de la Monte SM, Ho DD, Schooley RT, Hirsch MS, Richardson EP (1987) Subacute encephalomyelitis of AIDS and its relation to HTLV-III infection. Neurology 37: 562–569
Dewhurst S, Bresser J, Stevenson M, Sakai K, Evinger-Hodges MJ, Volsky DJ (1987a) Susceptibility of human glial cells to infection with HIV. Fed Eur Biochem Soc 213: 138–143
Dewhurst S, Sakai K, Bressner J, Stevenson M, Evinger-Hodges MJ, Volsky DJ (1987b) Persistent productive infection of human glial cells by human immunodeficiency virus (HIV) and by infectious molecular clones of HIV. J Virol 61: 3774–3782
Dewhurst S, Sakai K, Zhang XH, Wasiak A, Volsky DJ (1988) Establishment of human glial cell lines chronically infected with the human immunodeficiency virus. Virol 162: 151–159
Dickson DW (1986) Multinucleated giant cells in acquired immunodeficiency syndrome encephalopathy: origin from endogenous microglia? Arch Pathol Lab Med 110: 967–968
Elder G, Sever JL (1988) Neurologic disorders associated with AIDS retroviral infection. Rev Infect Dis 10: 286–302
Epstein LG, Sharer LR, Cho-ES, Meyenhofer M, Navia B, Price RW (1984-85) HTLV-III/LAV-like retrovirus particles in the brains of patients with AIDS encephalopathy. AIDS Res 1: 447–454
Epstein LG, Goudsmit J, Paul DA, Morrison SH, Connor EM, Oleske JM, Holland B (1987) Expression of human immunodeficiency virus in cerebrospinal fluid of children with progressive encephalopathy. Ann Neurol 21: 397–401
Evans LA, McHugh TM, Stites DP, Levy JA (1987) Differential ability of human immunodeficiency virus isolates to productively infect human cells. J Immunol 138: 3415–3418
Fisher AG, Ensoli B, Looney D, Rose A, Gallo RC, Saag MS, Shaw GM, Hahn BH, Wong-Staal F (1988) Biologically diverse molecular variants within a single HIV-isolate. Nature 334: 444–447
Gabuzda DH, Hirsch MS (1987) Neurological manifestation of infection with human immunodeficiency virus. Ann Intern Med 107: 383–391
Gabuzda DH, Ho DD, de la Monte SM, Hirsch MS, Rota TR, Sobel RA (1986) Immunohistochemical identification of HTLV-III antigen in brains of patients with AIDS. Ann Neurol 20: 289–295
Gartner S, Markovits P, Markovitz DM, Betts RF, Popovic M (1986a) Virus isolation from and identification of HTLV-III/LAV-producing cells in brain tissue from a patient with AIDS. JAMA 256: 2365–2371
Gartner S, Markovits P, Markovitz DM, Kaplan MH, Gallo RC, Popovic M (1986b) The role of mononuclear phagocytes in HTLV III/LAV infection. Science 233: 215–219
Gendelman HE, Narayan O, Kennedy-Stoskopf S, Kennedy PGE, Ghotbi Z, Clements JE, Stanley J, Pezeshkpour G (1986) Tropism of sheep lentiviruses for monocytes: susceptibility to infection and virus gene expression increase during maturation of monocytes to macrophages. J Virol 58: 67–74
Gendelman H, Orenstein JM, Martin M, Ferrua C, Mitra R, Phipps T, Wall L, Lane HC, Fauci AS, Burke DS, Skillman D, Meltzer MS (1988a) Efficient isolation and propagation of human immunodeficiency virus on recombinant colony-stimulating factor I-treated monocytes. J Exp Med 167: 1428–1441
Gendelman H, Meltzer MS, Van Arnold J, Burgess SK (1988b) HIV-mediated neurotoxicity overcome by trophic factors secreted by monocytes. 4th International conference on AIDS, Stockholm, Sweden. Abstract no 2060
Goudsmit J, de Wolf F, Paul DA, Epstein LG, Lange JMA, Krone WJA, Speelman H, Wolter EC, Van Der Noordaa J, Oleske JM, Van Der Helm HJ, Coutinho RA (1986) Expression of human immunodeficiency virus antigen (HIV-Ag) in serum and cerebrospinal fluid during acute and chronic infection. Lancet ii: 177–180
Gyorkey F, Melnick JL, Sinkovics JG, Gyorkey P (1985) Retrovirus resembling HTLV in macrophages in patients with AIDS. Lancet i: 106

Gyorkey F, Melnick JL, Gyorkey P (1987) Human immunodeficiency virus in brain biopsies of patients with AIDS and progressive encephalopathy. J Infect Dis 155: 870–876

Hahn BH, Gonda MA, Shaw GM, Popovic M, Hoxie JA, Gallo RC, Wong-Staal F (1985) Genomic diversity of the acquired immune deficiency syndrome virus HTLV-III. Different viruses exhibit greatest divergence in their envelope genes. Proc Natl Acad Sci USA 82: 4813–4817

Hahn BH, Shaw GM, Taylor ME, Redfield RR, Markham PD, Salahudden SZ, Wong-Staal F, Gallo RC, Parks ES, Parks WP (1986) Genetic variation in HTLV-III/LAV over time in patients with AIDS or at risk for AIDS. Science 232: 1548–1553

Harada S, Koyanagi Y, Yamamoto N (1985) Infection of HTLV-III/LAV in HTLV-1-carrying cells MT-2 and MT-4 and application in a plaque assay. Science 229: 563–566

Harouse JM, Kim J, McLaughlin M, Hoxie JA, Trojanowski JQ, Gonzalez-Scarano F (1988) HIV infection of neural cells is not blocked by OKT4. 4th International AIDS conference, Stockholm, Sweden. Abstract no 2625

Harper ME, Marselle LM, Gallo RC, Wong-Staal F (1986) Detection of lymphocytes expressing human T-lymphotropic virus type III in lymph nodes and peripheral blood from infected individuals by in situ hybridization. Proc Natl Acad Sci USA 83: 772–776

Hasse AT (1986) Pathogenesis of lentivirus infections. Nature 322: 130–136

Ho DD, Sarngadharan MG, Resnick L, di Marzo-Veronese F, Rota TR, Hirsch MS (1985a) Primary human T-lymphotropic virus type III infection. Ann Intern Med 103: 880–883

Ho DD, Rota TR, Schooley RT, Kaplan JC, Allan JD, Groopman JE, Resnick L, Felsenstein D, Andrews CA, Hirsch MS (1985b) Isolation of HTLV-III from cerebrospinal fluid and neural tissues of patients with neurologic syndromes related to the acquired immunodeficiency syndrome. N Engl J Med 313: 1493–1497

Ho DD, Rota TR, Hirsch MS (1986) Infection of monocyte/macrophages by human T lymphotropic virus type III. J Clin Invest 77: 1712–1715

Hollander H, Levy JA (1987) Neurologic abnormalities and recovery of human immunodeficiency virus from cerebrospinal fluid. Ann Intern Med 106: 692–695

Hoxie JA, Aloers JD, Rackowski JL, Huebner K, Haggarty BS, Cedarbaum AJ, Reed JC (1986) Alterations in T4 (CD4) protein and mRNA synthesis in cell infected with HIV. Science 234: 1123–1127

Klatzmann D, Champagne E, Chamaret S, Gruest J, Guetard D, Hercend T, Gluckman JC, Montagnier L (1984) T-lymphocyte T4 molecule behaves as the receptor for human retrovirus LAV. Nature 312: 767–768

Koenig S, Gendelman HE, Orenstein JM, Dal Canto MC, Pezeshkpour GH, Yungbluth M, Janotta F, Aksamit A, Martin MA, Fauci AS (1986) Detection of AIDS virus in macrophages in brain tissue from AIDS patients with encephalopathy. Science 233: 1089–1093

Koyanagi Y, Miles S, Mitsuyasu RT, Merrill JE, Vinters HV, Chen ISY (1987) Dual infection of the central nervous system by AIDS viruses with distinct cellular tropisms. Science 236: 819–822

Lee MR, Ho DD, Gurney ME (1987) Functional interaction and partial homology between human immunodeficiency virus and neuroleukin. Science 237: 1047–1051

Levy JA (1988) Mysteries of HIV: challenges for therapy and prevention. Nature 333: 519–522

Levy JA, Shimabukuro J, Hollander H, Mills J, Kaminsky L (1985) Isolation of AIDS-associated retroviruses from cerebrospinal fluid and brain of patients with neurological symptoms. Lancet ii: 586–588

Levy JA, Evans LA, Cheng-Mayer C, Pan L-Z, Lane A, Staben C, Dina D, Wiley C, Nelson J (1987) Biologic and molecular properties of the AIDS-associated retrovirus that affect antiviral therapy. Ann Inst Pasteur 138: 101–111

Levy RM, Bredesen DE, Rosenblum ML (1985) Neurological manifestations of the acquired immunodeficiency syndrome (AIDS): experience at UCSF and review of literature. J Neurosurg 62: 475–588

Lifson J, Coutre S, Huang E, Engelman E (1986a) Role of envelope glycoprotein carbohydrate in human immunodeficiency virus (HIV) infectivity and virus-induced cell fusion. J Exp Med 164: 2101–2106

Lifson JD, Feinberg MB, Reyes GR (1986b) Induction of CD4-dependent cell fusion by the HTLV-III/LAV envelope glycoprotein. Nature 323: 725–728

Lifson JD, Reyes GR, McGrath MS, Stein BS, Engleman EG (1986c) AIDS retrovirus induced cytopathology: giant cell formation and involvement of CD4 antigen. Science 232: 1123–1127

Lipkin WI, Parry G, Kiprov D, Abrams D (1985) Inflammatory neuropathy in homosexual men with lymphadenopathy. Neurology 35: 1479–1483

Madden PJ, Dalgleish AG, McDougal JS, Clapham PR, Weiss RA, Axel R (1986) The T4 gene encodes the AIDS virus receptor and is expressed in the immune system and the brain. Cell 47: 333–348

McDougal JS, Kennedy MS, Sligh JM, Cort SP, Mawle A, Nicholson JKA (1986) Binding of HTLV III/LAV to T4$^+$ T cells by a complex of the 110K viral protein and the T4 molecule. Science 231: 382–385

Meyenhofer MF, Epstein LG, Cho E-S, Sharer LR (1987) Ultrastructural morphology and intracellular production of human immmunodeficiency virus (HIV) in brain. J Neuropathol Exp Neurol 46: 474

Narayan O, Sheffer D, Clements JE, Tennekoon G (1985) Restricted replication of lentiviruses. Visna viruses induce a unique interferon during interaction between lymphocytes and infected macrophages. J Exp Med 162: 1954–1959

Navia BA, Price RW (1987) The acquired immunodeficiency syndrome dementia complex as the presenting or sole manifestation of human immunodeficiency virus infection. Arch Neurol 44: 65–69

Navia BA, Jordan BD, Price RW (1986a) The AIDS dementia complex. I. Clinical features. Ann Neurol 19: 517–524

Navia BA, Cho E-S, Petito CK, Price RW (1986b) The AIDS dementia complex. II. Neuropathology. Ann Neurol 19: 525–535

Nicholson JKA, Cross GD, Callaway CS, McDougal JS (1986) *In vitro* infection of human monocytes with human T-lymphotropic virus type III/lymphadenopathy-associated virus (HTLV III/LAV). J Immunol 137: 323–329

Nielson SL, Petito CK, Urmacher CD, Posner JB (1984) Subacute encephalitis in acquired immune deficiency syndrome: a postmortem study. Am J Clin Pathol 82: 678–682

Orenstein JM, Meltzer MS, Phipps T, Gendelman HE (1988) Cytoplasmic assembly and accumulation of human immunodeficiency virus types 1 and 2 in recombinant human colony-stimulating factor-1-treated human monocytes: an ultrastructural study. J Virol 62: 2578–2586

Petito CK, Navia BA, Cho E-S, Jordan BD, George DC, Price RW (1985) Vacuolar myelopathy pathologically resembling subacute combined degeneration in patients with acquired immunodeficiency syndrome. N Engl J Med 312: 874–879

Price R, Drew B, Sidtis J, Rosenblum M, Scheck A, Cleary P (1988) The brain in AIDS: central nervous system HIV-1 infection and AIDS dementia complex. Science 239: 586–592

Pumarola-Sune T, Navia BA, Cordon-Cardo C, Cho E-S, Price RW (1987) HIV antigens in the brains of patients with the AIDS dementia complex. Ann Neurol 21: 490–496

Resnick L, Di Marzo-Veronese F, Schupbach J, Tourtelottee WW, Ho DD, Muller F, Snapshak P, Vogt M, Groopman JE, Markham PD, Gallo RC (1985) Intra-blood-brain-barrier synthesis of HTLV-III-specific IgG in patients with neurologic symptoms associated with AIDS or AIDS-related complex. N Engl J Med 313: 1498–1504

Rosenblum ML, Levy RM, Bredesen DE (1988) AIDS and the nervous system. Raven, New York

Rutka JT, Giblin JR, Dougherty DV, McCullock JR, Rosenblum ML (1987) Establishment and characterization of five cell lines derived from human malignant gliomas. Acta Neuropathol (Berl) 75: 92–103

Saag MS, Hahn BH, Gibbons J, Li Y, Parks GS, Parks WP, Shaw GM (1988) Extensive variation of human immunodeficiency virus type-1 *in vivo*. Nature 334: 440–444

Salahuddin SZ, Rose RM, Groopman JE, Markham PD, Gallo RC (1986) Human T lymphotropic virus type III infection of human alveolar macrophages. Blood 68: 281–284

Sharer LR, Epstein LG, Cho E-S, Joshi VV, Meyenhofer MF, Rankin LF, Petito CK (1986) Pathologic features of AIDS encephalopathy in children: evidence for LAV/HTLV-III infection of brain. Hum Pathol 17: 271–284

Shaw GM, Harper ME, Hahn BH, Epstein LG, Gajdunsek DC, Price RW, Navia BA, Petito CK, O'Hara CJ, Groopman JE, Cho E-S, Oleske JM, Wong-Staal F, Gallo RC (1985) HTLV-III infection in brains of children and adults with AIDS encephalopathy. Science 227: 177–182

Snider WD, Simpson DM, Nielson S, Gold JWM, Metroka CE, Posner JB (1983) Neurologic complications of acquired immune deficiency syndrome: analysis of 50 patients. Ann Neurol 14: 403–418

Sodroski J, Goh WC, Rosen C, Campbell K, Haseltine WA (1986) Role of the HTLV-III/LAV envelope in syncytium formation and cytopathicity. Nature 322: 470–474

Sonnerborg AB, Ehrnst AC, Bergdahl SKM, Pehrson PO, Skoldenberg BR, Stennegard OO (1987) HIV isolation from cerebrospinal fluid in relation to immunological deficiency and neurological symptoms. AIDS Res 2: 89–93

Stoler MH, Eskin TA, Benn S, Angerer RC, Angerer LM (1986) Human T-cell lymphotropic virus type III infection of the central nervous system: a perliminary *in situ* analysis. JAMA 256: 2360–2364

Tateno M, Levy JA (1988) MT-4 plaque formation can distinguish cytopathic subtypes of the human immunodeficiency virus (HIV). Virology 167: 299–301

Tersmette M, Goede REY, Bert JM, Winkel IN, Gruters RA, Cuypers HT, Huisman HG, Miedema F (1988) Differential syncytium-induced capacity of human immunodeficiency virus isolates: frequent detection of syncytium-induced isolates in patients with acquired immune deficiency syndrome (AIDS) and AIDS-related complex. J Virol 62: 2026–2032

Vazeux R, Brousse N, Jarry A, Henin D, Marche C, Vedrene C, Mikol J, Wolff M, Michon C, Rozenbaum W, Bureau J-F, Montagnier L, Brahic M (1987) AIDS subacute encephalitis: identification of HIV-infected cells. Am J Pathol 126: 403–410

Weber JN, Stratton M, Weiss RA (1988) Infection of glial cells by HIV isolates: role of CD4 as receptor. 4th International conference on AIDS, Stockholm, Sweden. Abstract no 2623

Wiley CA, Schrier RD, Nelson JA, Lampert PW, Oldstone MBA (1986) Cellular localization of human immunodeficiency virus infection within the brains of acquired immune deficiency syndrome patients. Proc Natl Acad Sci USA 83: 7089–7093

Wong-Staal F, Shaw GM, Hahn B, Salahuddin SZ, Popovic M, Markham P, Redfield R, Gallo RC (1985) Genomic diversity of human T-lymphotropic virus type III (HTLV-III). Science 229: 759–762

Human Immunodeficiency Virus: Infection of the Nervous System

C. A. Wiley[1] and J. A. Nelson[2]

1 Introduction

Patients with acquired immune deficiency syndrome (AIDS) suffer from a variety of neurologic disorders (Levy et al. 1985). Between 10% and 25% of AIDS or AIDS-related complex (ARC) patients present with neurologic symptoms (Anders et al. 1986; Navia et al. 1986a). In the terminal phases of AIDS, up to one-half of all patients develop an encephalopathy, with autopsy studies demonstrating as much as 80% with CNS pathology (Levy et al. 1985; Anders et al. 1986). Even individuals who show only serologic evidence of human immunodeficiency virus (HIV) infection, who are otherwise clinically normal, develop neurologic illness (Mirra et al. 1986; Lenhardt et al. 1988). We will briefly review the peripheral neuromuscular and central encephalopathic illnesses associated with HIV infection and our understanding of the pathogenesis of these diseases.

[1] Department of Pathology (Neuropathology), University of California, San Diego, La Jolla, California 92093, USA
[2] Department of Immunology, Scripps Clinic and Research Foundation, La Jolla, California 92037, USA

Current Topics in Microbiology and Immunology, Vol. 160

2 Neuromuscular Disease Associated with HIV Infection

2.1 Muscle Disease

Neuromuscular disease is seen in 15%–20% of patients with AIDS. Usually this presents as a mild proximal muscle weakness; however, some patients develop severe muscular weakness with virtual paralysis and creatine kinase levels of up to 30 000 units (DALAKAS et al. 1986; DALAKAS and PEZESHKPOUR 1988). Despite striking clinical symptoms, when opportunistic infections (e.g., toxoplasmosis) are subtracted, the histopathologic changes seen in muscle biopsies are usually quite mild. Atrophic fibers, usually of the oxidative subtype, with minimal inflammation are noted along with rare necrotic fibers. The insensitivity of the histopathology might in part be explained by sampling problems, yet even when multiple biopsies are made from clinically involved muscles, the pathogenesis of weakness is frequently unexplained. Biopsies of some patients have demonstrated a rod myopathy (DALAKAS et al. 1987), showing interesting similarities to transgenic mice (see NERENBERG, this volume).

Numerous proposals have been made regarding the pathogenesis of the neuromuscular disease seen in AIDS patients. Given the wasting nature of AIDS, nutritional factors were considered. However, there is no correlation between nutritional status and presence of neuromuscular abnormalities, and some patients develop neuromuscular symptoms as the presenting symptom of AIDS before nutritional factors come into consideration. No exogenous toxins have been demonstrated in this diverse group of patients, but endogenous toxins (e.g., immune cytokines) have a potential role in neuromuscular dysfunction. Numerous infectious agents have been noted in the muscles of AIDS patients; however, there are no reports of direct viral infection of muscle fibers. HIV has been detected only in monocytes surrounding abnormal muscle fibers. In sum, there is no unifying theory regarding the pathogenesis of muscular disorders in AIDS. Given the variability of pathologic findings, it is assumed that no single mechanism prevails. The paucity of findings suggests that some muscular abnormalities could be secondary to a neural process.

2.2 Peripheral Nerve Disease

There has been increasing recognition of peripheral neuropathies in HIV-infected individuals (LIPKIN et al. 1985; CORNBLATH et al. 1987). Several different subtypes of peripheral neuropathy have been noted in up to 20% of AIDS patients (PARRY 1988; DALAKAS et al. 1988). The most frequently observed form is a distal symmetric polyneuropathy which runs a monophasic course and potentially responds to plasmapharesis (LIPKIN et al. 1985) or antiviral therapy (DALAKAS et al. 1988). Histologic studies have shown that in a group of 21 AIDS or ARC patients where only half had clinical symptomatology all but 1 showed pathology within peripheral nerves (DE LA MONTE et al. 1988). In three-fourths of these, peripheral nerve specimens showed moderate or servere demyelination, one-third showed axonal degeneration, and another third showed mononuclear inflammation. Like muscle disease, this variety of peripheral nerve findings does not suggest a single etiology.

Despite the frequently noted peripheral neuropathy, the rare reports of HIV isolation from peripheral nerve specimens (HO et al. 1985 and DE LA MONTE et al. 1988) are not convincing that this neuropathy is directly related to HIV-infection. In fact, systematic studies of peripheral nerves of AIDS patients uniformly fail to identify HIV but occasionally identify cytomegalovirus (CMV) (MORGELLO et al. 1987; MAH et al. 1987; GRAFE and WILEY 1989). This suggests that the peripheral neuropathy seen in AIDS patients is secondary to opportunistic CMV infection of peripheral nerve elements rather than direct HIV infection of the nerve. Peripheral nerve damage could be mediated by CMV in several different ways. Direct Schwann cell infection could lead to segmental demyelination with resultant conduction failure. Alternatively, local CMV infection and the resultant inflammatory response could lead to edema and intrinsic compression of the involved nerve with axonal degeneration.

3 Central Nervous System Disease Associated with HIV Infection

Neuropsychiatric symptoms are frequently a dominant clinical feature of AIDS (GRANT et al. 1987). Attendant generalized encephalopathy and myelopathy can occur at various times of illness (NAVIA et al. 1986a; ASHER et al. 1987). Neurologic symptoms range from impaired concentration and memory loss, to global cognitive dysfunction with severe motor impairment. These symptoms and signs of diffuse degeneration of the CNS in the most extreme forms lead to a complete vegetative state.

The most frequent clinical symptoms of the AIDS dementia complex are attributable to subcortical changes consistent with deep gray and white matter disease (NAVIA et al. 1986b). Motor abnormalities are consistent with a myelopathy due to interruption of corticospinal tracts. Laboratory analyses of CSF show fairly nonspecific changes of mild pleocytosis with slight increase in protein content. For the most part neuroradiologic studies have not proven to be as sensitive indicators of neurologic disease as is the clinical symptomatology. Computed tomography of the head has shown generalized cortical atrophy with sulcal widening and ventricular dilution (POST et al. 1986; LEVY RM et al. 1986). More recently magnetic resonance images (MRIs) have shown more frequent abnormalities particularly with T2-weighted images, which are sensitive to inflammatory changes in water content (i.e., edema). MRI has shown clinically unsuspected multifocal high-intensity signals in various regions of the nervous system. Abnormalities later correlated with a viral encephalitis appear as high-intensity signals in the basal ganglia and deep white matter (BELMAN et al. 1986).

In the first few years of the AIDS epidemic it was difficult to separate the variety of encephalopathic processes according to specific etiologies. Clinical studies have shown toxoplasmosis to be the most common cause of mass lesions and cryptococcossis to be the most common fungal infection. Nevertheless, nonfocal symptoms still provide significant problems in linking clinical encephalopathy with specific pathologic processes. In many cases, diagnosis requires a brain biopsy or diagnosis is deferred until autopsy, where approximately one-third of AIDS patients have a viral encephalopathy variably diagnosed as "subacute encephalitis" (NIELSEN et al. 1984; MOSKOWITZ et al. 1984; SNIDER et al. 1983). This encephalitis is particularly associated with patients suffering from a subcortical dementia termed "AIDS dementia complex" (NAVIA et al. 1986a).

The viral agent responsible for the encephalitis frequently is not apparent. Given the nature of the immunosuppression seen in AIDS patients, there are numerous candidate opportunistic viruses that could explain the encephalitis. Despite numerous associations of various agents with the AIDS dementia complex, the pathogenesis of this disease is still unknown. Neuropathologic examination has demonstrated histopathologic findings in the expected distribution of the clinical symptomatology (NAVIA et al. 1986b; SHARER et al. 1986). A spongiform change is noted in clinically and radiologically involved regions consistent with intra- or extracellular edema as expected by MRI. Involved regions are remarkable for a frequently minimal inflammatory response, warranting the term "unconventional" encephalopathy as that seen in slow viral infections. Inflammation is restricted to a few chronic perivascular infiltrates along with parenchymal macrophages. This type of viral encephalitis is in striking contrast to the intense acute and chronic inflammation characteristic of Herpes simplex virus (HSV) or visna virus infection of the brain.

Perhaps the most sensitive indicator of the chronic CNS damage is the rather atypical gliosis noted in the brains of AIDS dementia complex patients. The gliosis is atypical in that vacuolated regions show a diffuse immunocytochemical staining for glial fibrillary acidic protein (GFAP) while surrounding tissues show sharply demarcated GFAP-positive astrocytes (KLEIHUES et al. 1985). This gliosis is most pronounced in basal ganglia and deep white matter structures with a sharp demarcation from overlying uninvolved cortical structures. Accompanying the chronic perivascular inflammation is a calcific vasculopathy more frequently noted in children than adults (EPSTEIN et al. 1984–85). In severe cases, frank tissue destruction is noted along with multinucleated giant cells, particularly in the basal ganglia and deep white matter, but also noted in the spinal cord. Many of the multinucleated giant cells appear to be fused macrophage elements. While severe cases can show destruction of spinal cord tissue, the most frequent pathology noted in the cord consists of a vacuolar myelopathy (PETITO et al. 1985). This myelopathy most frequently involves the lateral columns where it is probably secondary to Wallerian degeneration of cortical tracts (GRAFE and WILEY, 1989). This explanation does not, however, account for the posterior column involvement. Damage of these ascending tracts implies damage to dorsal root ganglia or other peripheral components of the nervous system. In summary, the unique neuropathology suggests either a unique etiologic agent or a unique host response to a common agent. A wide range of potential viral agents is obvious.

3.1 Cellular Localization of HIV

In January 1985, SHAW et al. published a survey of tissues infected by HIV. They observed a remarkable quantitative distribution of HIV in various autopsy tissues. As expected, lymphoid organs contained chromosomally integrated and unintegrated HIV, but, surprisingly, so did the brain. In fact, in several cases the concentration of viral nucleic acids in the brain was greater than that in the spleen. This suggested that the CNS was a potential reservoir of the AIDS virus, protected within an immunologically privileged site. Not unexpectedly, this localization was considered as evidence of "neurotropism" of HIV and would at least superficially explain the neurologic symptoms associated with AIDS. This 1985 article launched an intensive histologic search of the CNS for HIV by many groups. HIV was readily isolated from CSF and nervous system tissues (LEVY et al. 1985; HO et al. 1985). Because of the absence of good reagents (both antibodies and nucleic acid probes) early histologic studies were slow. Ultrastructural studies demonstrated rare HIV particles within the CNS in cells of questionable origin (EPSTEIN et al. 1984–85; SHARER et al. 1986). We (WILEY et al. 1986) and others (GABUZDA et al. 1986; STOLER et al. 1986; KOENIG et al. 1986; WARD et al. 1987; VAZEUX et al. 1987; ROSTAD et al. 1987; PUMAROLA-SUNE et al. 1987) definitively localized HIV within macrophages and endothelial cells using immunocytochemistry and in situ hybridization (Fig. 1A, B). Table 1 summarizes the eight 1986 and 1987 published reports localizing HIV within the CNS. Using a variety of techniques all investigators identified HIV within macrophages and multinucleated giant cells. In selected cases most investigators found these two cells frequently infected; however, KOENIG et al. detected HIV in "rare" macrophages and only 5% of multinucleated giant cells. Half of the reports demonstrated endothelial cell infection, although all found these cells infected to a lesser degree than macrophages. Rare infection of glia and neurons were noted in exceptional cases. In our study of 93 autopsies performed from 1982 through 1987 we detected HIV in 34 cases. Our compilation of the relative frequency of infection of different cell types in these cases is shown in Table 2. Twenty percent of all AIDS autopsies showed frequently infected macrophages. Half of these cases also showed endothelial cell infection, although not to the same degree. Only one case showed frequent infection of CNS parenchymal cells. Recent evidence of fetal CNS infection by HIV (MARION et al. 1986; JOVAISES et al. 1985) requires more extensive corroboration and analysis to discern the lineage of infected cells.

In vivo localization of HIV has led to more questions than answers. During the 18-month hiatus between discovery of HIV within the CNS and cellular

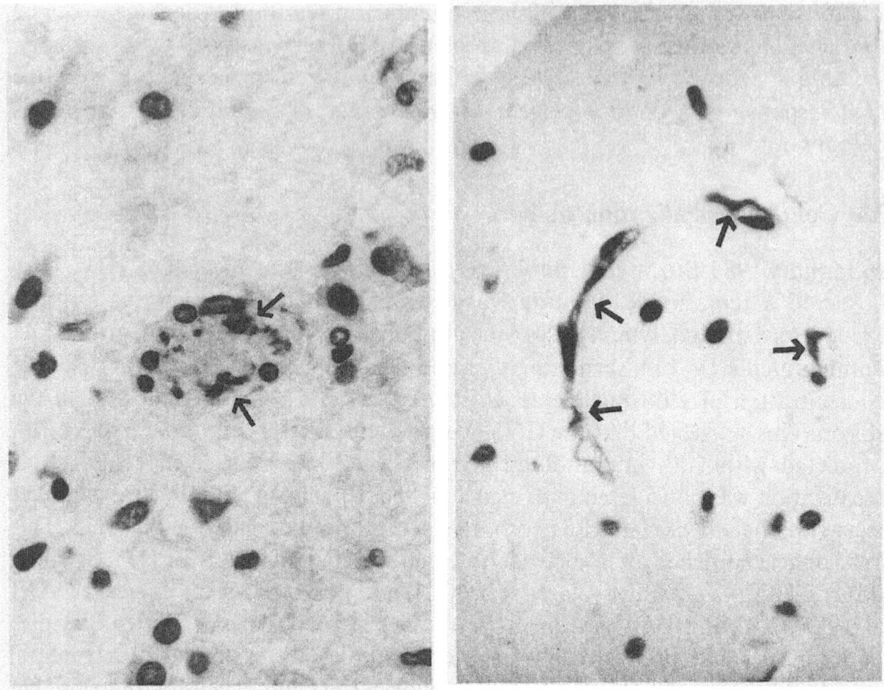

Fig. 1a. Paraffin section from the brain of an AIDS patient with HIV encephalopathy, immunostained for HIV gp41. A multinucleated giant cell contains abundant reaction product (*arrows*) while surrounding CNS parenchyma does not stain counterstained with hematoxylin, × 120, **b** Paraffin section from the brain of an AIDS patient with HIV encephalopathy, immunostained for HIV gp41. The cytoplasm of endothelial cells is stained with the immunoreaction product (*arrows*). Counter-stained with hematoxylin, × 120

localization of the virus, the label "neurotropic" was solidly affixed to HIV (SHAW et al. 1985; PRICE et al. 1988) despite the absence of significant demonstration of neuronal infection. It was, perhaps, only natural to suspect that neurologic symptoms would be due to direct neuronal infection. The congruence of pathologic and clinical findings in the face of a low-level neuronal and glial infection suggests an indirect cause of neuronal dysfunction.

A second question was raised when it was conceded that the amount of virus detected was incommensurate with the pathology. As ultrastructural studies have repeatedly shown (EPSTEIN et al. 1984/85; SHARER et al. 1986; KOENIG et al. 1986; WILEY et al. 1988), virion is exquisitely rare, and to date has only been definitively detected within multinucleated giant cells and macrophages. Viral antigen is more frequently observed but does not come close to approximating the histopathology. In our laboratory and others (VAZEUX et al. 1987), viral nucleic acids are detected no more frequently than viral antigens. It had been concluded that insensitivities in techniques or inadequate preservation must explain the discrepancy between paucity of virus and abundance of pathology.

Table 1. Published localization of HIV within the brains of autopsied AIDS patients

Reference	Neurons	Glia	Endos	Macros	MNGCs	Pts + /total	Immunocytochemistry	NA hybridization
Wiley et al. 1986	+ (1/12)	+ (1/12)	+ (7/12)	+	+	9/12	MM gp41	YES
Gabuzda et al. 1986	–	–	+	+	+	5/13	Goat poly	ND
Stoler et al. 1986	+ (Rare)	+	?	+	+	2	ND	YES
Koenig et al. 1986	–	–	–	+ (Rare)	+ (5%)	2	ND	YES
Ward et al. 1987	–	+	+	+	+	?/13	Rabbit poly	ND
Vazeux et al. 1987	–	–	?	+	+	5/16	MMp25/gp110	YES
Rostad et al. 1987	–	–	+	+	+	4/20	ND	YES
Pumarola-Sune et al. 1987	+	+ (4/8)	?	+	+	?/8	MM p25	ND

Numbers in parentheses indicate frequency of cases with pertinent observation. Percentages in parentheses indicate frequency of cells in individual cases. ND, not done; NA hybridization, in situ hybridization; Immunocytochemistry, immunocytochemistry using either polyclonal antisera or mouse monoclonal antibodies (MMSs), to specified viral peptides; Pts + /total, ratio of selected patients with HIV detected compared with total sample size

Table 2. Immunohistologic studies of the brains of 34 AIDS autopsies in which HIV was detected. Double-label immunocytochemistry or cytomorphology was used to identify the lineage of cells containing HIV antigens. HIV-infected macrophages were frequently detected in half of these cases, while infected endothelial cells were detected in one-third of the cases. Lineage of some occasionally infected cells could not be discerned in all cases

	Rare	Occasional	Frequent
Macrophages	7	12	17 (20%)
Endothelia	1	12	8 (10%)
Other ?	–	13	1 (1%)

3.2 Time Course of HIV Infection

There are now several reports of documented aseptic meningitis with HIV seroconversion (i.e., primary infection) (COOPER et al. 1985; CARNE et al. 1985; GOUDSMIT et al. 1986), indicating that CNS infection by HIV can be the primary and perhaps only symptom of HIV infection. Virus can be frequently recovered from CSF, and intrathecal synthesis of antibodies to HIV is good evidence of chronic intrathecal infection by HIV (HO et al. 1985; GOUDSMIT et al. 1987). It is inappropriate, however, to jump from this clear evidence of meningeal infection to conclusions regarding the infection of the CNS parenchyma. These two types of infection do not necessarily go hand in hand. In fact as exemplified by HSV, viral meningitis and encephalitis are usually completely dissociated. Primary infection by HSV is frequently accompanied by an aseptic meningitis, but seldom, except in congenital infections, accompanied by an encephalitis. Rather, HSV encephalitis occurs during a period of reactivation of a latent infection with intraaxonal passage of the virus into the CNS (presumably through the olfactory nerve). Given the frequency of recovery of HIV from the CSF, the infrequency of identification of HIV within the CNS parenchyma, and the topologic distribution of parenchymal infection, HIV probably does not spread from the meninges into the CNS. What prevents HIV or HSV from progressing from the leptomeninges into the CNS is unclear, but may relate to viral or host factors (see below).

In most cases of HIV encephalitis, the amount of tissue damage appears incommensurate with the amount of virus detected. We have recently shown this to be particularly true in children with HIV encephalitis (WILEY et al. in press). Children showing a progressive clinical course have HIV within macrophages and endothelial cells in the CNS. Children showing a plateau clinical course do not have detectable HIV antigens within the CNS despite the presence of tissue damage in a distribution similar to the children with a progressive course. This difference in presence of HIV antigen within damaged brains of children with AIDS suggests that either; children with a plateau course have mounted an effective immune response leading to clearance of the virus from the CNS after initial damage, or CNS damage occurs prior to the appearance of HIV within the CNS parenchyma that is later colonized by HIV-infected cells.

One problem with the adult and pediatric clinical studies to date is that they are examining a late stage of infection that might not represent what occurs at the initiation of CNS infection. The first cell infected by HIV can only be surmised from these fragmentary data. For CNS infection to occur virus must first breach the blood-brain barrier. In analogy with CMV, we would postulate early HIV infection of endothelial cells with damage to nervous system tissue followed by preferential productive growth within macrophages elements that migrate into the CNS. In patients with CMV encephalitis or retinitis, one identifies abundant productive infection of neuronal elements and rare infection of endothelial cell elements. Recently, we and others have documented CMV endothelial cell infection (WILEY and NELSON 1988). With such a seeding of the CNS endothelium, CMV could invade an environment where it would undergo unrestricted growth.

3.3 Presence of Other Viral Infections

As mentioned above, several other viruses have been identified within the brains of AIDS patients. As seen in other immune-compromised patients, individuals with AIDS are afflicted with progressive multifocal leukoencephalopathy (PML). PML in AIDS patients (GRAY et al. 1987; WILEY et al. 1988) demonstrates the same distribution of JC viral nucleic acids as previously described in other immunosuppressed patients. PML is a remarkable encephalitis because of the limited inflammatory response restricted to macrophage infiltration alone. In AIDS patients with longstanding PML, many of these macrophages fuse into multinucleated giant cells. This fusion is presumably mediated by macrophage infection with HIV (see below). HSV has also been found within the CNS of AIDS patients, but it is a rare complication, especially when considered in the setting of the frequency of systemic HSV infection. There are no pathologic data to suggest that HSV is any more neurovirulent in AIDS patients than it is in immunologically intact individuals.

On the other hand, cytomegalovirus infection is very commonly observed in the brains of AIDS patients (MORGELLO et al. 1987). In our series 60% of AIDS patients with a subacute encephalitis are infected by CMV (WILEY and NELSON 1988). Dissemination of CMV to the CNS is characterized by two distinct histopathologic patterns: microscopic infarctions and microglial nodules. Presence of CMV within microinfarcts could be explained by hematogenous infection of CNS endothelial cells followed by swelling of the cell with occlusion of the lumen and infarction of dependent brain tissue, or by secondary CMV colonization of previously infarcted regions. Microglial nodules associated with CMV infection of the CNS are frequently seen in other immunosuppressed patients (e.g., organ transplant recipients) as well as AIDS patients. The pathogenic mechanism of this form of diffused dissemination is unclear, but requires a breach of the normal blood-brain barrier with penetration through the endothelial cell, surrounding basement membrane and astrocytic foot processes. Once within the CNS, CMV is capable of productively infecting most CNS cells and progressing by direct extension from a central nidus.

We have noted a high incidence of individual brains coinfected by HIV and another virus, particularly CMV. As would be the case in any of numerous other opportunistic infections, coinfection of the brain by HIV could be the result of inflammatory processes within the CNS that result in the influx of latently HIV infected monocytic elements. This would be particularly important in PML where the inflammatory process is notable for an almost exclusive macrophage infiltration. In fact, some of the first cases where HIV was detected within the CNS were cases of PML. Cytomegalovirus acts in a manner similar to other opportunistic infections by attracting latently HIV infected monocytes to the CNS. However, our recent observation of coinfection of individual cells by HIV and CMV implies another level of cooperativity (NELSON et al. 1988). Infection by one virus might change host cell permissiveness to infection by a second virus (e.g., CMV infection of neuronal or glial cells permitting HIV infection, or HIV infection of macrophages and endothelial cells permitting productive CMV infection). Alternatively, infection of cells permissive to both viruses (e.g., macrophages) might lead to synergistic intracellular interactions with preferential expression of one virus. Such an interaction is known to occur in vivo for adenovirus and SV 40 (RABSON et al. 1964) or a defective parvovirus (JANIK et al. 1981), and has been suggested in vitro for HIV and herpes viruses (GENDELMAN et al. 1986; MOSCA et al. 1987). In the latter case immediate early genes of the herpes virus can act as transactivators of HIV.

4 Potential Mechanisms of HIV-Mediated CNS Damage

Despite the marked amount of heterogeneity in clinical symptoms, there is usually good correlation with clinically involved regions and neuropathologic evidence of damage. The distribution of damage would be consistent with other described subcortical dementias (e.g., Huntington's and Parkinson's dementia). Unfortunately, the level of tissue destruction is incommensurate with the amount of virus detected. Either there is more virus present than detected, or damage is an indirect effect of virus infection.

There are several reasons to expect that we are not detecting all virus present. Working with human material introduces variations in preservation and degradation of antigens and nucleic acids. Since there is no control over the time points examined, our picture of the neuropathologic processes is composed of random snapshots. However, in animal models of retroviral infections (e.g., visna; NARAYAN and CORK 1985) where control over preservation and time points of sampling is complete, there is also little evidence of virus. An alternate explanation to the absence of significant levels of the virus would be to assume that the probes used are too insensitive. Despite enzyme amplification with antibody probes and highly radioactive nucleic acid probes, an infection of neuronal or glial cells at very low copy number could lead to neuronal dysfunction without detectable virus. In all other viral/host systems studied with these techniques, cellular dysfunction has only been noted where virus could be detected. Perhaps future in situ experiments using the polymerase chain reaction technique might identify previously undetected neuronal or glial HIV infection.

If our assays are as sensitive as we suspect, then the nervous system damage by HIV must be amplified through an indirect mechanism. These indirect effects could be targeted against either neuronal soma or their unmyelinated and myelinated axonal processes. The indirect effects could be mediated by multiple mechanisms. Retroviral proteins may be toxic to cells. Recently the envelope protein of HIV I has shown homology to the cytokine neuroleukin (GURNEY et al. 1986; LEE et al. 1987). By actively competing for the neuroleukin neuronal receptor, GP 120 has been shown to decrease neuronal survival in vitro. If such an indirect effect occurs in vivo, this could account for neuronal dysfunction in the face of limited viral antigen. However, this hypothesis has been brought into question with the recent discovery that neuroleukin is probably a normal brain enzyme, a phosphohexose isomerase (CHAPUT et al. 1988; FAIK et al. 1988).

Simple vasogenic edema due to chronic encephalitis with disruption of the blood-brain barrier also would be expected to lead to swelling and conduction dysfunction. This nonspecific damage could account for both clinical and pathologic findings. A more specific and yet indirect effect of HIV infection could be mediated by blood-brain barrier disruption due to endothelial cell infection. Compromise of endothelial function in conjunction with a severe systemic infection and the numerous associated toxic serologic factors might lead to CNS damage.

An alternate mechanism for amplifying viral damage of the CNS could be mediated by the immune system. Oligoclonal immunoglobulin bands can be detected in the CSF of patients with AIDS encephalopathy (HO et al. 1985; RESNICK et al. 1985; GOUDSMIT et al. 1987). Analysis of the concentration of CSF immunoglobulin is consistent with intrathecal synthesis. Many of these immunoglobulins are directed against HIV envelope and gag proteins gp41 and p24. However, these immunoglobulins might also play a role in mediating CNS damage through binding to infected cells or through molecular mimicry by binding to normal nerve cell membranes. Either systemic or locally produced cytokines are likely to appear within HIV-infected brains. Infected and uninfected macrophages have the potential for producing numerous potent cytokines including interferon, interleukin I, and tumor necrosis factor, all of which could amplify viral damage and directly affect CNS function. Recently, astrocytes have also been shown to secrete a tumor necrosis-like factor which is glial toxic and could therefore mediate white matter damage (ROBBINS et al. 1987).

5 Heterogeneity in CNS Infection

It is difficult to explain why this purportedly "neurotropic" (PRICE et al. 1988) virus causes CNS disease in only a subset of patients. Given the high viremia and frequent recovery of HIV from the CSF, it is not readily apparent why certain patients develop HIV encephalopathy and others do not. One theory to explain this would propose that most people infected by HIV will be afflicted with CNS disease if they survive a long enough period to permit viral entry into the CNS. While it is true that there is an increased incidence of CNS disease with increased duration of disease, it is also true that some AIDS patients present with "subacute encephalitis" as an initial symptom of their disease. These data would suggest that some factor unique to the virus or host is responsible for the presence or absence of HIV encephalopathy.

5.1 Neurotropic Viral Strain Theory of CNS Disease

Another theory particularly in vogue today is that there is a specific viral strain ("neurotropic strain") that is responsible for CNS disease. This theory draws strength from analogy with reovirus where a strain-specific sigma 1 protein confers a receptor-defined neurotropism (TYLER et al. 1983). The neurotropic HIV strain theory proposes that certain individuals are either initially infected with, or eventually evolve through intraindividual antigenic drift, strains of the virus that are neurotropic (i.e., infect nervous system cells). In fact, there are now several studies that document different properties of HIV strains isolated from the CNS versus those isolated from the peripheral blood. The CNS isolates grow particularly well in macrophages, while those from the blood grow well in T-cell lines (KOYANGI et al. 1987). In fact, this finding is not surprising given the observation that much of the HIV present within the brain is present in macrophages. However, attempts to date to show unique nucleic acid sequences specific to "neurotropic" strains of HIV have been unrewarding. Historically, studies with lymphochoriomeningitis virus, HSV, and CMV strains have indicated that mutations in many regions can effect pathogenesis and no single region controls neurotropism.

5.2 Defective Host Immunity Theory of CNS Disease

An alternate theory to account for the presence of HIV encephalopathy within certain individuals would suggest that there is something unique in the host that accounts for the predilection toward CNS disease. There are several possible host attributes that could account for this predilection. Severe immune compromise in certain hosts might be one such attribute that could permit higher viral titers and greater invasion of the virus into normally protected areas like the CNS. While this has not been examined systematically, there is no apparent correlation between severity of immune suppression (T-cell ratios, etc.) and presence or absence of CNS disease. The fact that some patients exhibiting little immune compromise develop CNS disease would suggest that a more specific attribute of the host must account for the presence or absence of neurologic disease. Such a specific attribute might lie within the CNS or immune system of afflicted individuals.

6 Summary

Cynics would say it has taken the scientific community a long time to achieve very little progress in our understanding of HIV-mediated CNS damage. We cannot yet say with surity how neuronal function is affected. However, when viewed through the perspective that retroviral diseases of the human nervous system are newly recognized diseases, significant progress has been made in the 3 years since HIV infection was noted within the CNS. We have a lot to learn about *how* retroviruses damage the CNS, but atleast the questions are better defined.

References

Anders KH, Guerra WF, Tomiyasu U, Verity MA, Vinters HV (1986) The neuropathology of AIDS. Am J Pathol 124: 537–558

Asher DM, Epstein LG, Goudsmit J (1987) Human immunodeficiency virus in the central nervous system In: Gottlieb MS et al. (eds) Current topics in AIDS, vol 1. Wiley, pp 225–246

Belman AL, Lantos G, Horoupian D, Novick BE, Ultmann MH, Dickson DW, Rubinstein A (1986) Calcification of the basal ganglia in infants and children. Neurology 36: 1192–1199

Carne CA, Teder RS, Smith A, Sutherland S, Elkington SG, Daly HM, Preston FE, Craske J (1985) Acute encephalopathy coincident with seroconversion for anti-HTLV-III. Lancet i: 1206–1208

Chaput M, Claes V, Portetelle D, Cludts I, Cravador A, Burny A, Gras H, Tartar A, (1988) The neurotropic factor neuroleukin is 90% homologous with phosphohexose isomerase. Nature 332: 454–455

Cooper DA, Gold J, MacLean P, Donovan B, Finlayson R, Michelmore HM, Brooke P, Penny R (1985) Acute AIDS retrovirus infection. Definition of a clinical illness associated with seroconversion. Lancet i: 537–540

Cornblath DR (1988) Treatment of the neuromuscular complications of human immunodeficiency virus infection. Ann Neurol 23 (Suppl): S88–S91

Cornblath DR, McArthur JC, Kennedy PGE, Witte AS, Griffin JW (1987) Inflammatory demyelinating peripheral neuropathies associated with human T-cell lymphotropic virus type III infection. Ann Neurol 21: 32–40

Dalakas MC, Pezeshkpour GH, Gravell M, Sever JL (1986) Polymyositis associated with AIDS retrovirus. JAMA 256: 2381–2383

Dalakas MC, Pezeshkpour GH (1988) Neuromuscular disease associated with human immunodeficiency virus infection. Ann Neurol 23 (Suppl): S38–S48

Dalakas MC, Pezeshkpour GH, Flaherty M (1987) Progressive nemaline (rod) myopathy associated with HIV infection. N Engl J Med 317: 1602–1603
Dalakas MC, Yarchoan R, Spitzer R, Elder G, Sever JL (1988) Treatment of human immunodeficiency virus-related polyneuropathy with 3'-azido-2', 3'-dideoxythymidine. Ann Neurol 23 (Suppl): S92–S94
de la Monte SM, Gabuzda DH, Ho DD, Brown RH, Hedley-Whyte ET, Schooley RT, Hirsch MS, Bahn AK (1988) Peripheral neuropathy in the acquired immunodeficiency syndrome. Ann Neurol 23: 485–492
Epstein LG, Sharer LR, Cho E-S, Myenhofer M (1984/85) HTLV-III/LAV-like retrovirus particles in the brains of patients with AIDS encephalopathy. AIDS Res 1: 447–454
Faik P, Walker JIG, Redmill AM, Morgan MJ (1988) Mouse glucose-6-phosphate isomerase and neuroleukin have identical 3' sequences. Nature 322: 455–456
Gabuzda DH, Ho DD, de la Monte SM, Hirsch MS, Rota TR, Sobel RA (1986) Immunohistochemical identification of HTLV-III antigen in brains of patients with AIDS. Ann Neurol 20: 289–295
Gartner S, Markovits P, Markovit DM, Betts RF, Popovic M (1986) Virus isolation from and identification of HTLV-III/LAV-producing cells in brain tissue from a patient with AIDS. JAMA 256: 2365–2371
Gendelman HE, Phelps W, Feigenbaum L, Ostrove JM, Adachi A, Howley PM, Khoury G, Ginsberg HS, Martin MA (1986) *Trans*-activation of human immunodeficiency virus long terminal repeat sequences by DNA viruses. Proc Natl Acad Sci USA 83: 9759
Goudsmit J, de Wolf F, Paul DA, Epstein LG, Lange JMA, Krone WJA, Speelman H, Wolters ECH, van der Noordaa J, Oleske JM, van der Helm HJ, Countinho RA (1986) Expression of human immunodeficiency virus antigen (HIV-Ag) in serum and cerebrospinal fluid during acute and chronic infection. Lancet ii: 177–180
Goudsmit J, Epstein LG, Paul DA, van der Helm JH, Dawson GJ, Asher DM, Yanagihara R, Wolff AV, Gibbs CJ, Gajdusek DC (1987) Intra-blood-brain barrier synthesis of human immunodeficiency virus antigen and antibody in humans and chimpanzees. Proc Natl Acad Sci USA 84: 3876–3880
Grafe MR, Wiley CA (1989) Spinal cord and peripheral nerve pathology in AIDS: the roles of cytomegalovirus and human immunodeficiency virus. Ann Neurol 25: 561–566
Grant I, Atkinson J, Hesselink JR, Kennedy CJ, Richman DD, Spector SA, McCutchan JA (1987) Evidence for early central nervous system involvement in the acquired immunodeficiency syndrome (AIDS) and other human immunodeficiency virus (HIV) infections. Ann Intern Med 107: 828–836
Gray F, Gherardi R, Baudrimont M, Gaulard P, Meyrignac C, Vedrenne C, Poirier J (1987) Leucoencephalopathy with multinucleated giant cells containing human immune deficiency virus-like particles and multiple opportunistic cerebral infections in one patient with AIDS. Acta Neuropathol (Berl) 73: 99–104
Gurney ME, Heinrich SP, Lee MR, Yin H-S (1986) Molecular cloning and expression of neuroleukin, a neurotropic factor for spinal and sensory neurons. Science 234: 566–581
Ho D, Rota TR, Schooley RT, Kaplan JC, Allan JD, Groopman JE, Resnick L, Felsenstein D, Andrews CA, Hirsch MS (1985) Isolation of HTLV-III from cerebrospinal fluid and neural tissue of patients with neurologic syndromes related to the acquired immunodeficiency syndrome. N Engl J Med 313: 1493–1497
Janik JE, Huston MM, Rose JA (1981) Location of adenovirus genes required for the replication of adenovirus-associated virus. Proc Natl Acad Sci USA 78: 1925–1929
Jovaisas E, Koch MA, Schafer A, Stauber M, Lowenthal D (1985) LAV/HTLV-III in 20-week fetus. Lancet ii: 1129
Kleihues P, Lang W, Burge PC, Budka H, Vogt M, Maurer R, Luthy R, Siegenthaler W (1985) Progressive diffuse leukoencephalopathy in patients with acquired immune deficiency syndrome (AIDS). Acta Neuropathol (Berl) 68: 333–339
Koenig S, Gendelman HE, Orenstein JM, Dal Canto MC, Pezeshkpour GH, Yungbluth M, Janotta F, Aksamit A, Martin MA, Fauci AS (1986) Detection of AIDS virus in macrophages in brain tissue from AIDS patients with encephalopathy. Science 233: 1089–1093
Lee MR, Ho DD, Gurney ME (1987) Functional interaction and partial homology between human immunodeficiency virus and neuroleukin. Science 237: 1047–1051
Lenhardt TM, Super MA, Wiley C (1988) Severe central nervous system gliosis in an asymptomatic HIV seropositive individual. Ann Neurol 23: 209–210
Levy JA, Shimabukuro J, Hollander H, Mills J, Kaminsky L (1985) Isolation of AIDS-associated

retroviruses from cerebrospinal fluid and brain of patients with nevrologic symptoms. Lancet ii: 586–588
Levy RM, Bredesen DE, Rosenblum ML (1985) Neurological manifestations of the acquired immunodeficiency syndrome (AIDS): experience at UCSF and review of the literature. J Neurosurg 62: 475–495
Levy RM, Rosenbloom S, Perrett LV (1986) Neuroradiologic findings in AIDS: a review of 200 cases. Am J Pathol 147: 977–983
Lipkin WI, Parry G, Kiprov D, Abrams D (1985) Inflammatory neuropathy in homosexual men with lymphadenopathy. Neurology 35: 1479–1483
Mah V, Vartavarian L, Akers M-A, Vinters HV (1987) Abnormalities of peripheral nerve in patients with acquired immune deficiency syndrome (AIDS). J Neuropathol Exp Neurol 46: 380
Marion RW, Wiznia AA, Hutcheon RG, Rubinstein A (1986) Human T-cell lymphotropic virus III (HTLV-III) embryopathy. Am J Dis Child 140: 638–640
Mirra SZ, Anand R, Spira TJ (1986) HTLV-III/LAV infection of the central nervous system in a 57-year-old man with progressive dementia of unknown cause. N Engl J Med 314: 1191–1192
Morgello S, Cho E-S, Nielsen S, Devinsky O, Petito CK (1987) Cytomegalovirus encephalitis in patients with acquired immunodeficiency syndrome: an autopsy study of 30 cases and a review of the literature. Hum Pathol 18: 289–297
Mosca JD, Bednarik DP, Raj NBK, Rosen CA, Sodroski JG, Haseltine WA, Pitha PM (1987) Herpes simplex virus type-1 can reactivate transcription of latent human immunodeficiency virus. Nature 325: 67
Moskowitz LB, Hensley GT, Chan JC, Gregorios J, Conley FK (1984) The neuropathology of acquired immune deficiency syndrome. Arch Pathol Lab Med 108: 867–872
Narayan O, Cork LC (1985) Lentiviral diseases of sheep and goats: chronic pneumonia leukoencephalomyelitis and arthritis. Rev Infect Dis 7: 89–98
Navia BA, Jordan BD, Price RW (1986a) The AIDS dementia complex: I. Clinical features. Ann Neurol 19: 517–524
Navia BA, Cho E-S, Petito CK, Price RW (1986b) The AIDS dementia complex: II. Neuropathology. Ann Neurol 19: 525–535
Nelson JA, Reynolds-Kohler C, Oldstone MBA, Wiley CA (1988) HIV and HCMV coinfect brain cells in patients with AIDS. Virology 165: 286–290
Nielsen SL, Petito CK, Urmacher CD, Posner JB (1984) Subacute encephalitis in acquired immune deficiency syndrome: a postmortem study. Am J Clin Pathol 82: 678–682
Parry GJ (1988) Peripheral neuropathies associated with human immunodeficiency virus infection. Ann Neurol 23 (Suppl): S49–S53
Petito CK, Navia BA, Cho E-S, Jordan BD, George DC, Price RW (1985) Vacuolar myelopathy pathologically resembling subacute combined degeneration in patients with the acquired immunodeficiency syndrome. N Engl J Med 312: 874–879
Post MJD, Sheldon JJ, Hensley GT, Soila K, Tobias JA, Chan JC, Quencer RM, Moskowitz LB (1986) Central nervous system disease in acquired immunodeficiency syndrome: prospective correlation using CT, MR imaging, and pathologic studies. Radiology 158: 141–148
Price RW, Brew B, Sidtis J, Rosenblum M, Scheck AC, Cleary P (1988) The brain in AIDS: central nervous system HIV-I infection and AIDS dementia complex. Science 239: 586–592
Pumarola-Sune T, Navia BA, Cordon-Cardo C, Cho E-S, Price RW (1987) HIV antigen in the brains of patients with AIDS dementia complex. Ann Neurol 21: 490–496
Rabson AS, O'Connor GT, Berezesky IK, Paul FJ (1964) Enhancement of adenovirus growth in African green monkey kidney cell cultures by SV40. Proc Soc Exp Biol Med 116: 187–190
Resnick L, Marzo-Veronese F, Schupbach J, Tourtellotte WW, HO DD, Muller F, Shapshak P, Vogt M, Groopman JE, Markham PD, Gallo RC (1985) Intra blood-brain barrier synthesis of HTLV-III specific IgG in patients with neurologic symptoms associated with AIDS or AIDS-related complex. N Engl J Med 313: 1498–1504
Robbins DS, Shirazi Y, Drysdale B-E, Lieberman A, Shin HS, Shin ML (1987) Production of cytotoxic factor for oligodendrocytes by stimulated astrocytes. J Immunol 139: 2593–2597
Rostad SW, Sumi SM, Shaw CM, Olson K, McDougall JK (1987) Human immunodeficiency virus (HIV) infection in brains with AIDS-related leukoencephalopathy. AIDS Res Hum Retrovirus 3: 4
Sharer LR, Epstein LG, Cho E-S, Joshi VV, Meyenhofer MF, Rankin LF, Petito CK (1986) Pathologic features of AIDS encephalopathy in children: evidence for LAV/HTLV-III infection of brain. Hum Pathol 17: 217–284

Shaw GM, Harper GM, Hahn BH, Epsterin LG, Gajdusek DC, Price RW, Navia BA, Petito CK, O'Hara CJ, Groopman JE, Cho E-S, Oleske JM, Wong-Staal F, Gallo RC (1985) HTLV-III infection in brains of children and adults with AIDS encephalopathy. Science 227: 177–182
Snider WD, Simpson DM, Nielsen S, Gold JWM, Metroka CE, Posner JB (1983) Neurological complications of acquired immune deficiency syndrome: analysis of 50 patients. Ann Neurol 14: 403–418
Stoler MH, Eskin TA, Benn S, Angerer RC, Angerer LM (1986) Human T-cell lymphotropic virus type III infection of the central nervous system. JAMA 256: 2360–2364
Tyler KL, Bronson RT, Byers KB, Fields B (1985) Molecular basis of viral neurotropism: experimental reovirus infection. Neurology 35: 88–92
Vazeux R, Brousse N, Jarry A, Henin D, Marche C, Vedrenne C, Mikol J, Wolff M, Michon C, Rozenbaum W, Bureau J-F, Montagnier L, Brahic M (1987) AIDS subacute encephalitis: identification of HIV-infected cells. Am J Pathol 126: 403–410
Ward JM. O'Leary TJ, Baskin GB, Benveniste R, Harris CA, Nara PL, Rhodes RH (1987) Immunohistochemical localization of human and simian immunodeficiency viral antigens in fixed tissue sections. Am J Pathol 127: 199–205
Wiley CA, Belman Al, Dickson D, Rubinstein A, Nelson JA (in press) Human immunodeficiency virus within the brains of children with AIDS. Clin Neuropath
Wiley CA, Nelson JA (1988) Role of human immunodeficiency virus and cytomegalovirus in AIDS encephalitis. Am J Pathol 133: 73–81
Wiley CA, Schrier RD, Nelson JA, Lampert PW, Oldstone MBA (1986) Cellular localization of human immunodeficiency virus infection within the brains of acquired immune deficiency syndrome patients. Proc Natl Acad Sci USA 83: 7089–7093
Wiley CA, Grafe M, Kennedy C, Nelson JA (1988) HIV and JC virus in AIDS patients with progressive multifocal leukoencephalopathy. Acta neurol (Berl) 76: 338–346

Subject Index

Preface

Although retroviruses have long been associated with a variety of animal diseases, active research in the field of human retroviruses dates from the discovery of human immunodeficiency virus (HIV) in association with acquired immunodeficiency syndrome (AIDS). The enormous research efforts in this field have been directed toward understanding the nature of the virus and toward its elimination through preventive vaccination and the cure of the disease.

Human T-cell leukemia virus (HTLV-1) was the first member of the human retrovirus family to be discovered. It was implicated as the cause of adult T-cell leukemia (ATL) even before the association of HIV and AIDS was established. Research on HTLV-1 has, however, been lagging behind that of HIV because of the importance of AIDS. Today HTLV-1 and possibly closely related HTLV-2 are associated with a variety of human neurologic diseases, and research activities in this field may show that human retroviruses can cause a variety of human diseases in addition to those affecting the nervous system.

Papers in this volume attempt to acquaint the reader with the present state of research into retrovirus infection and related diseases of the nervous system.

Prominent among diseases of experimental and domestic animals caused by retroviruses are demyelinative disorders, arthritic and rheumatic illnesses, and progressive motor neuron disorders. The wild mouse ecotropic virus causes a disease that is discussed in the chapters by GARDNER and PORTIS. GARDNER first described the disease in wild ("outbred") mouse populations and he has contributed to defining the susceptibility/resistance genes that control this infection. PORTIS has been studying the disorder in inbred populations and mapping the viral gene(s) involved. The study of wild mouse ecotropic virus demonstrates how a retrovirus functions in a genetically defined and manipulable experimental animal to cause slow, progressive disease and points up the similarities

Vol. 137: **Mock, Beverly; Potter, Michael (Ed.):** Genetics of Immunological Diseases. 1988. 88 figs. XI, 335 pp. ISBN 3-540-19253-0

Vol. 138: **Goebel, Werner (Ed.):** Intracellular Bacteria. 1988. 18 figs. IX, 179 pp. ISBN 3-540-50001-4

Vol. 139: **Clarke, Adrienne E.; Wilson, Ian A. (Ed.):** Carbohydrate-Protein Interaction. 1988. 35 figs. IX, 152 pp. ISBN 3-540-19378-2

Vol. 140: **Podack, Eckhard R. (Ed.):** Cytotoxic Effector Mechanisms. 1989. 24 figs. VIII, 126 pp. ISBN 3-540-50057-X

Vol. 141: **Potter, Michael; Melchers, Fritz (Ed.):** Mechanisms in B-Cell Neoplasia 1988. Workshop at the National Cancer Institute, National Institutes of Health, Bethesda, MD, USA, March 23–25, 1988. 1988. 122 figs. XIV, 340 pp. ISBN 3-540-50212-2

Vol. 142: **Schüpbach, Jörg:** Human Retrovirology. Facts and Concepts. 1989. 24 figs. 115 pp. ISBN 3-540-50455-9

Vol. 143: **Haase, Ashley T.; Oldstone, Michael B. A. (Ed.):** In Situ Hybridization. 1989. 33 figs. XII, 90 pp. ISBN 3-540-50761-2

Vol. 144: **Knippers, Rolf; Levine, A. J. (Ed.):** Transforming Proteins of DNA Tumor Viruses. 1989. 85 figs. XIV, 300 pp. ISBN 3-540-50909-7

Vol. 145: **Oldstone, Michael B. A. (Ed.):** Molecular Mimicry. Cross-Reactivity between Microbes and Host Proteins as a Cause of Autoimmunity. 1989. 28 figs. VII, 141 pp. ISBN 3-540-50929-1

Vol. 146: **Mestecky, Jiri; McGhee, Jerry (Ed.):** New Strategies for Oral Immunization. International Symposium at the University of Alabama at Birmingham and Molecular Engineering Associates, Inc. Birmingham, AL, USA, March 21–22, 1988. 1989. 22 figs. IX, 237 pp. ISBN 3-540-50841-4

Vol. 147: **Vogt, Peter K. (Ed.):** Oncogenes. Selected Reviews. 1989. 8 figs. VII, 172 pp. ISBN 3-540-51050-8

Vol. 148: **Vogt, Peter K. (Ed.):** Oncogenes and Retroviruses. Selected Reviews. 1989. XII, 134 pp. ISBN 3-540-51051-6

Vol. 149: **Shen-Ong, Grace L. C.; Potter, Michael; Copeland, Neal G. (Ed.):** Mechanisms in Myeloid Tumorigenesis. Workshop at the National Cancer Institute, National Institutes of Health, Bethesda, MD, USA, March 22, 1988. 1989. 42 figs. X, 172 pp. ISBN 3-540-50968-2

Vol. 150: **Jann, Klaus; Jann, Barbara (Ed.):** Bacterial Capsules. 1989. 33 figs. XII, 176 pp. ISBN 3-540-51049-4

Vol. 151: **Jann, Klaus; Jann, Barbara (Ed.):** Bacterial Adhesins. 1990. 23 figs. XII, 192 pp. ISBN 3-540-51052-4

Vol. 152: **Bosma, Melvin J.; Phillips, Robert A.; Schuler, Walter (Ed.):** The Scid Mouse. Characterization and Potential Uses. EMBO Workshop held at the Basel Institute for Immunology, Basel, Switzerland, February 20–22, 1989. 1989. 72 figs. XII, 263 pp. ISBN 3-540-51512-7

Vol. 153: **Lambris, John D. (Ed.):** The Third Component of Complement. Chemistry and Biology. 1989. 38 figs. X, 251 pp. ISBN 3-540-51513-5

Vol. 154: **McDougall, James K. (Ed.):** Cytomegaloviruses. 1990. 58 figs. IX, 286 pp. ISBN 3-540-51514-3

Vol. 155: **Kaufmann, Stefan H. E. (Ed.):** T-Cell Paradigms in Parasitic and Bacterial Infections. 1990. 24 figs. IX, 162 pp. ISBN 3-540-51515-1

Vol. 156: **Dyrberg, Thomas (Ed.):** The Role of Viruses and the Immune System in Diabetes Mellitus. 1990. 15 figs. XI, 142 pp. ISBN 3-540-51918-1

Vol 157: **Swanstrom, Ronald; Vogt, Peter K. (Ed.):** Retroviruses. Strategies of Replication. 1990. 40 figs. XII, 260 pp. ISBN 3-540-51895-9

Vol 158: **Muzyczka, Nicholas (Ed.):** Viral Expression Vectors. 1990. approx. 20 figs. approx. XII, 190 pp. ISBN 3-540-52431-2

Vol 159: **Gray, David; Sprent, Jonathan (Ed.):** Immunological Memory. 1990. 38 figs. XII, 156 pp. ISBN 3-540-51921-1

GPSR Compliance

The European Union's (EU) General Product Safety Regulation (GPSR) is a set of rules that requires consumer products to be safe and our obligations to ensure this.

If you have any concerns about our products, you can contact us on ProductSafety@springernature.com

In case Publisher is established outside the EU, the EU authorized representative is:

Springer Nature Customer Service Center GmbH
Europaplatz 3
69115 Heidelberg, Germany

Batch number: 10620878

Printed by Printforce, the Netherlands